Modern Computational Approaches To Traditional Chinese Medicine

Modern Computational Approaches To Traditional Chinese Medicine

Zhaohui Wu
Huajun Chen
Xiaohong Jiang

Zhejiang University,
Hangzhou, China

AMSTERDAM • BOSTON • HEIDELBERG • LONDON • NEW YORK • OXFORD
PARIS • SAN DIEGO • SAN FRANCISCO • SINGAPORE • SYDNEY • TOKYO

Elsevier
32 Jamestown Road, London NW1 7BY
225 Wyman Street, Waltham, MA 02451, USA

British Library Cataloguing-in-Publication Data
A catalogue record for this book is available from the British Library

Library of Congress Cataloging-in-Publication Data
A catalog record for this book is available from the Library of Congress

ISBN: 978-0-323-28272-7

For information on all Elsevier publications
visit our website at store.elsevier.com

This book has been manufactured using Print On Demand technology. Each copy is produced to order and is limited to black ink. The online version of this book will show colour figures where appropriate.

Contents

Preface

Traditional Chinese medicine (TCM) is both an ancient and a living medical system using fully developed theoretical and practical ideas. In China, traditional medicine accounts for around 40% of all health care delivered. TCM is recognized as an essential component of Chinese culture, and the preservation and modernization of this Chinese cultural heritage is prioritized in the Chinese government's planning program.

TCM has an independently evolving knowledge system, which is expressed mainly in the Chinese language. TCM knowledge discovery and knowledge management have emerged as innovative approaches for the preservation and utilization of this knowledge system. It aims at the computerization of TCM information and knowledge to provide intelligent resources and supporting evidence for clinical decision making, drug discovery, and education.

Specifically, the expansion of TCM practice results in the ongoing accumulation of more and more research documents and clinical data. The major concern in TCM is how to consolidate and integrate the data, and enable efficient retrieval and discovery of novel knowledge from the massive data. Typically, this requires an interdisciplinary approach involving Chinese culture, modern health care, and life sciences. For example, in order to map a global network of herb–drug interactions revealing drug communities, explicit knowledge should be integrated from a plurality of heterogeneous data resources in health care and the life science domain, including electronic health records, literature databases, and domain knowledge databases.

Additionally, TCM knowledge is commonly available in the form of ancient classics and confidential family records, which are disparate among people and organizations across geographical areas. Novel knowledge integration and discovery approaches are thus required to link data across database and organizational boundaries so as to enable more intuitive queries, search, and navigation without the awareness of these boundaries.

The goal of exploring effective methods for knowledge discovery and management in TCM in this book is to provide a systematic interface which can bridge the linguistic gap, cultural gap, and methodological gap between TCM and Western science, by extracting intelligent resources from physicians' theoretical and practical knowledge and applying computational approaches to promote the automatic progress of clinical decision making, drug discovery, and education.

This book compiles a number of recent research results from the Traditional Chinese Medicine Informatics Group of Zhejiang University. This book reports systematic approaches for developing knowledge discovery and knowledge management applications in TCM. These approaches feature in the utilization of the modern Semantic Web and data mining methods for more advanced data integration, data analysis, and integrative knowledge discovery. Driven by the heterogeneous distribution and obscure literature of TCM data, these methods and techniques mentioned in this book aim to analyze and understand such huge amounts of data in a controllable style and turn this into integrated knowledge. The digested knowledge could thus be used to promote systemized knowledge discovery and mining, so that TCM experts, physicians, even ordinary people, can gain excellent experience of effective knowledge acquisition. For example, a large-scale TCM domain ontology is utilized to improve the quality of search and query, and to interpret statistically important patterns in those reported approaches. Semantic graph mining methodology is developed for discovering interesting patterns from a large and complex network of medical concepts. The platform and underlying methodology has proved effective in cases such as personalized health care with TCM characteristics, TCM drug discovery, and safety analysis. This book can be a reference book for researchers in TCM informatics. Generally, the topics in this book cover major fundamental research issues, track current challenges, and present core applications. Specifically, we mainly make important contributions to TCM knowledge discovery and management from the following aspects:

1. TCM Data Mining

TCM is a completely dependent discipline, and is a complementary knowledge system to modern biomedical science. Due to diverse and increasing biomedical data, it is difficult to obtain effective information for applications from such massive data. Different forms of TCM data also hinder integration of information sources from different disciplines. Data mining techniques provide flexible approaches to uncovering implicit relationships in these data sources.

Chapter 2 assumes that related genes of the same syndrome will have some biological functional relationships, and thus constitute a functional gene network. We generated syndrome-based gene networks from 200,000 syndrome-gene relations, in order to analyze the functional knowledge of genes from the syndrome perspective. The primary results suggest that it is worthy of further investigation. In Chapter 3, a path-finding algorithm in the context of complex networks was designed to detect network motifs. Performance evaluation is made to learn about the data block size, node number, and network bandwidth in considering the MapReduce-based path-finding performance. In addition to the architectural level of data mining methods, Chapter 5 presents a unified Domain-driven Data Mining platform, which carries out data mining applications by resource orchestration through web services from a variety of heterogeneous intelligence resources and data. The effectiveness of this platform has been proved by a series of in-use applications in the TCM domain. Semantic associations are complex relationships between resource entities, which is a topic studied in Chapter 10 to explore and interpret the knowledge assets of TCM. A case study is demonstrated that discovers and integrates relationships and interactions on TCM herbs from distributed data sources. Chapter 13 presents a novel approach, which utilizes node and link types together with the topology of a semantic graph to derive a similarity graph from linked datasets. Semantic similarity is calculated through semantic similarity transition during the process of generating a similarity graph.

2. TCM Knowledge Discovery and Retrieval

Confronted with the increasing popularity of TCM and the huge volume of TCM data, there is an urgent need to explore these sources effectively so as to generate useful knowledge, by the techniques of knowledge discovery and retrieval. Knowledge discovery is one proper methodology for analyzing such heterogeneous data.

Chapter 1 provides readers with a perfect overview of knowledge discovery in TCM, including knowledge discovery in a database (KDD) for the research of Chinese medical formula, Chinese herbal medicine, TCM syndrome research, and TCM clinical diagnosis. Chapter 4 attempts to investigate data-quality issues particularly in the field of TCM. Three data-quality aspects are highlighted as key dimensions, including representation granularity, representation consistency, and completeness, and practical methods and techniques are proposed to handle data-quality problems. In order to achieve seamless and interoperable e-Science for TCM, Chapter 6 presents a comprehensive approach to building dynamic and extendable e-Science applications for information integration and service coordination of TCM. The semantic e-Science infrastructure uses domain ontologies to integrate TCM database resources and services, and delivers a semantic experience with browsing, searching, querying, and knowledge discovery for users. Chapter 11 introduces an in-use application deployed at the China Academy of Traditional Chinese Medicine (CATCM), in which over 70 legacy relational databases are semantically interconnected by a shared ontology, providing semantic query, search, and navigation services to the TCM communities. Chapter 12 proposes a probability-based semantic relationship discovery method, which combines a TCM domain ontology and more than 40,000 relative publications so as to uncover hidden semantic relationships between resources. A probabilistic RDF model is defined and used to store semantic relations identified as uncertain and assigned with a probability.

3. TCM Knowledge Modeling

Knowledge representation is a primary step in understanding the nature of diverse domains and conducting useful applications, especially in scientific fields such as biology, economics, and medicine. A lot of knowledge modeling techniques are proposed to solve different levels of knowledge representation obstacles so as to construct an applicable infrastructure.

As a complete knowledge system, TCM researches into human health care via a different approach compared to orthodox medicine. In Chapter 7, a unified traditional Chinese medical language system (UTCMLS) is developed through an ontology approach, which will support TCM language knowledge storage, concept-based information retrieval, and information integration. It is a huge project which was collaborated on by 16 distributed groups. Moreover, unlike Western Medicine, knowledge in TCM is based on inherent rules or patterns, which can be considered as causal links. Chapter 8 presents a semantic approach to building a TCM knowledge model with the capability of rule reasoning using OWL 2, a kind of web ontology language defined by the W3C consortium. The knowledge model especially focuses on causal relations among syndromes and symptoms and changes between syndromes. The evaluation results suggest that the approach clearly displayed the causal relations in TCM and shows great potential in TCM knowledge mining. The on-demand and scalability requirement ontology-based systems should go beyond the use of static ontology and be able to self-evolve and specialize in the domain knowledge. Chapter 9 refers to the context-specific portions from large-scale ontologies like TCM ontology as sub-ontologies. A sub-ontology evolution approach is proposed based on a genetic algorithm for reusing large-scale ontologies.

For a short overview, the book is specifically organized as follows: Chapter 1 gives an overview of the progress of knowledge discovery in TCM; Chapter 2 introduces a specific text mining application that integrates TCM literature and MEDLINE for functional gene networks analysis; Chapter 3 introduces a novel approach that utilizes a MapReduce framework to improve mining performance with an application of network motif detection for TCM; Chapter 4 discusses the data-quality issue for knowledge discovery in TCM; Chapter 5 reports on a service-oriented mining engine and several case studies from TCM; Chapter 6 elaborates on a systematic approach to TCM knowledge management based on Semantic Web technology; Chapter 7 introduces a large-scale ontology effort for TCM and describes the unified traditional Chinese medical language system; Chapter 8 reports an approach to modeling causal knowledge for TCM using OWL 2; Chapter 9 discusses the ontology evolution issue as related to TCM web ontology; Chapter 10 proposes an ontology-based technical framework for hypothesis-driven Semantic Association Mining, which allows a knowledge network to emerge through the communication of Semantic Associations by a multitude of agents in terms of hypotheses and evidence; Chapter 11 describes a Semantic Web approach to knowledge integration for TCM; Chapter 12 introduces a probabilistic approach to discovering semantic relations from large-scale traditional Chinese medical literature; Chapter 13 presents methods of analyzing semantic linked data for TCM.

This book is the result of years of study, research, and development of the faculties, Ph.D. candidates, and many others affiliated to the CCNT Lab of Zhejiang University. We would like to give particular thanks to Xuezhong Zhou, Peiqin Gu, Xiangyu Zhang, Yuxin Mao, Xiaoqing Zheng, Yi Feng, Yu Zhang, Chunyin Zhou, Tong Yu, Jinhua Mi, Yang Liu, Junjian Jian, Sen Liu, Mingkui Liu, Hao Shen, Jinhuo Tao, and many others who have devoted their energy and enthusiasm to this book and relevant projects.

We would also like to give particular thanks to our long-term collaborator: the China Academy of Chinese Medical Science (CACMS). We would like to thank Hongxin Cao, director of CACMS, and Baoyan Liu, vice director of CACMS, who gave us ongoing strong support over the past 10 years. Also, we are grateful to Meng Cui, the director of the Institute of TCM informatics of CACMS, and all of his kind colleagues. Without their strong support, this book could not have been finished.

In addition, the work in this book was mainly sponsored by the "973" Program (National Basic Research Program of China) of the Semantic Grid initiative (No. 2003CB317006); the National Science Fund for the Distinguished Young Scholars of China NSF Program (No. NSFC60533040); and the Program for New Century Excellent Talents in University of the Ministry of Education of China (No. NCET-04-0545). The work was also partially supported by the National Program for Modern Service Industry (No. 2006BAH02401); "863" Program (National High-Tech Research and Development Program of China) (Nos. 2006AA01A122, 2009AA011903, 2008AA01Z141); the Program for Changjiang Scholar (IRT0652); the NSFC Programs under Grant No. NSFC61070156, NSFC60873224, Important Programs of Zhejiang Sci-Tech Plan (No. 2008C03007).

Zhaohui Wu
Zhejiang University, Hangzhou, China
November 2011

List of Contributors

Huajun Chen
William K. Cheung
Weiyu Fan
Yi Feng
Peiqin Gu
Xiaohong Jiang
Baoyan Liu
Yang Liu
Bin Lu
Jun Ma
Yuin Mao
Yuxin Mao
Jinhua Mi
Zhongkai Mi
Jinmin Tang
Wenya Tian
Heng Wang
Yimin Wang
Lancheng Wu
Zhaohui Wu
Aining Yin
Tong Yu
Ruen Zhang
Xiangyu Zhang
Xiaogang Zhang
Chunyin Zhou
Xuezhong Zhou
Zhongmei Zhou

1 Overview of Knowledge Discovery in Traditional Chinese Medicine[1]

1.1 Introduction

As a complete medical knowledge system other than orthodox medicine, traditional Chinese medicine (TCM) has played an indispensable role in health care for Chinese people for thousands of years. The holistic and systematic ideas of TCM are essentially different from the thinking modes based on reductionism in Western medicine. With the development of modern science, people came to realize the limitations of reductionism and began to lay more emphasis on systematic thinking patterns, such as systems biology [1]. Based on the methodology of holism, TCM plays a unique role in advancing the development of life science and medicine. Meanwhile, with the dramatic increase in the prevalence of chronic conditions, chemical medicines cannot totally satisfy the needs of health maintenance, disease prevention, and treatment. Human health demands the large-scale development and application of natural medicines, to which TCM experiences and knowledge can contribute a lot. The ever-increasing use of Chinese herbal medicine (CHM) and acupuncture worldwide is a good indication of the public interest in TCM [2–6].

Countless TCM practices and theoretical research over thousands of years accumulated a great deal of knowledge in the form of ancient books and literature. In China, the domestic collection of ancient books about TCM published before the Xinhai Revolution (1911) reaches 130,000 volumes. Besides, thousands of studies on TCM treatments are published yearly in journals all around the world. There were more than 600,000 journal articles during the period 1984–2005. With such a vast volume of TCM data, there is an urgent need to use these precious resources effectively and sufficiently. Besides, the last decade has been marked by unprecedented growth in both the production of biomedical data and the amount of published literature discussing it. Thus, it is an opportunity, but also a pressing need, to connect TCM with modern life science.

Knowledge discovery in databases (KDD) is one proper methodology to analyze and understand such huge amounts of data. As an interdisciplinary area between artificial intelligence, databases, statistics, and machine learning, the idea of KDD

[1] Reprinted from Feng Y, Wu Z, Zhou X, Zhou Z, Fan W. Knowledge discovery in traditional Chinese medicine: state of the art and perspectives. Artif Intell Med 2006;38:219–36. © 2006 Elsevier B.V., with permission from Elsevier.

Modern Computational Approaches To Traditional Chinese Medicine. DOI: http://dx.doi.org/10.1016/B978-0-12-398510-1.00001-7

came into being in the late 1980s. The most prominent definition of KDD was proposed by Fayyad et al. [7] in 1996. In that paper, KDD was defined as "the non-trivial process of identifying valid, novel, potentially useful, and ultimately understandable patterns in data." This definition may also be applied to "data mining" (DM). Indeed, in the recent literature of DM and KDD, the terms are often used interchangeably or without distinction. However, according to classical KDD methodologies [7], DM is the knowledge extraction step in the KDD process, which also involves the selection and preprocessing of appropriate data from various sources and proper interpretation of the mining results. Typical DM methods include concept description, association rule mining, classification and prediction, clustering analysis, time-series analysis, text mining, and so forth [8]. During the last two decades, the field of KDD has attracted considerable interest in numerous disciplines, ranging from telecommunications, banking, and marketing to scientific analysis. It is also the case within medical environments. The discipline of medicine deals with complex organisms, processes, and relations, and KDD methodology is particularly suitable to handle such complexity [9]. Besides, the advent of computer-based patient records (CPRs) and data warehouses contribute greatly to the availability of medical data and offer voluminous data resources for KDD. Also, the need to increase medical knowledge of human beings pushes researchers to carry out knowledge discovery not only in CPRs and clinical warehouses but also in biomedical literature databases. The creation of new medical knowledge with DM techniques is listed as one of the 10 grand challenges of medicine by Altman [10]. As Roddick et al. [11] indicate, the application of KDD to medical datasets is a rewarding and highly challenging area. Due to the ever-increasing accumulation of biomedical data and the pressing demand to explore these resources, the methods of knowledge discovery have been widely applied to analyze medical information over the decades. Reviews of KDD in the medical area from different perspectives can be found in Refs. [9,11–16]. However, the topic of knowledge discovery in TCM (KDTCM) is not covered in these reviews.

Considering the fast-growing number of researches carried out on KDTCM, it is also necessary and helpful to provide an overview of recent KDTCM research. As a complementary medical system, TCM is quite different from Western medicine, both in practice and in theory. In view of the high domain-specificity of KDD technology, it is more necessary to gain an insight into KDTCM. Motivated by these needs, this chapter focuses on the introduction and summarization of existing work about KDTCM. Because a great amount of KDTCM work is reported only in Chinese literature, the literature search is conducted in both English and Chinese publications, and the major KDTCM studies published there are covered in this review. For each work, the KDD methods used in the study are introduced, as well as corresponding results. In particular, some studies with interesting results are highlighted, such as novel TCM paired drugs discovered by frequent itemset analysis, the laboratory-confirmed relationship between CRF gene and kidney YangXu syndrome discovered by text mining, the high proportion of toxic plants in the botanical family Ranunculaceae discovered by statistical analysis, and the association between the M-cholinoceptor blocking drug and Solanaceae discovered by association rule mining. The existing work in KDTCM demonstrates that the usage of KDD in TCM is both feasible and promising. Meanwhile, it should be noted that the TCM field is still nearly a piece of virgin soil with copious amounts of hidden gold as far as KDD methodology is concerned. To ease gold mining in this field, the future directions of KDTCM research are also provided in this article based on a discussion of existing work.

The rest of this chapter is arranged as follows. The prerequisite for applying KDD is the digitalization of the vast amount of data. Thus, an overview of currently available TCM data resources is first presented in Section 1.2. Subsequently, the review of KDTCM work is presented in four research subfields in Section 1.3, including KDD for the research of Chinese medical formulae (CMF), KDD for the research of CHM, KDD for TCM syndrome research, and KDD for TCM clinical diagnosis. Based on a discussion of these KDTCM studies, the current state and main problems of KDTCM work in each subfield are summarized in Section 1.4, and the future directions for each subfield are also presented. Finally, we conclude in Section 1.5.

1.2 The State of the Art of TCM Data Resources

Data availability is the first consideration before any knowledge discovery task can be undertaken. In this section, we introduce the current state of TCM data resources, especially those data resources focusing on TCM in particular.

As a significant part of complementary and alternative medicine (CAM), literature reporting TCM issues can be found in the main CAM databases, such as CAM on PubMed (Complementary and Alternative Medicine subset of PubMed), AMED (Allied and Complementary Medicine Database), CISCOM (Centralized Information Service for Complementary Medicine), and CAMPAIN (Complementary and Alternative Medicine and Pain Database). A more comprehensive list of TCM databases can be found in Ref. [17]. Currently, the primary data resources specific to TCM include China TCM Patent Database (CTCMPD) [18], TradiMed Database [19], TCM chemical database [20], and TCM-online Database System [21]. CTCMPD has been established by Patent Data Research and Development Center, a subsidiary of the Intellectual Property Publishing House of the State Intellectual Property Office (SIPO) of China. More than 19,000 patent records and over 40,000 TCM formulae published from 1985 to the present are contained in CTCMPD [18]. TradiMed Database was built by the Natural Product Research Institute at Seoul National University, Republic of Korea. Based on various Chinese and Korean medical classics, TradiMed represents a combination of traditional medicine knowledge and modern medicine. So far, TradiMed contains information of 3199 herbs, 11,810 formulae, 20,012 chemical compositions of herbs, and 4080 diseases [19]. The TCM chemical database was developed by the National Key Laboratory of Bio-chemical Engineering at the Institute of Process Engineering, Chinese Academy of Sciences. This database contains detailed information of 9000 chemicals isolated from nearly 4000 natural sources used in TCM and provides in-depth bioactivity data for many of the compounds [20].

In this section, we place our emphasis on the TCM-online Database System. To the best of our knowledge, currently the TCM-online Database System is the largest TCM data collection in the world. The prototype of TCM-online was first built in the late 1990s. In 1998, the advanCed Computing aNd sysTem (CCNT) Lab in the College of Computer Science in Zhejiang University and China Academy of Traditional Chinese Medicine (CATCM) began to collaborate in building the scientific databases for TCM and established a unified web-accessible multidatabase query system TCMMDB [21] that integrates 17 branches in the whole country. Through the input from nearly 300 scientists from more than 30 colleges, universities, and academies of TCM, this system has already integrated more than 50 databases, including the Traditional Chinese Medical Literature Analysis and Retrieval System (TCMLARS), Traditional Chinese Drug Database (TCDBASE), and Database of Chinese Medical Formula. TCMMDB was replaced by the Grid-based system TCM-Grid [22] in 2002, which provides more powerful functions, such as dynamic registration, binding, and associated navigation. The TCM-Grid system was further extended to a semantic-based database Grid named DartGrid in 2002. At present, these databases are available as the TCM-online Database System via web site [23] and CD-ROM versions. Besides, a large-scale ontology-based Unified TCM Language System (UTCMLS) [24] has been developed to support concept-based information retrieval and information integration since 2001. All these efforts help to realize the organization, storage, and sharing of TCM data, which provide a feasible environment for the effective implementation of KDD technology.

Today, the TCM-online Database System integrates more than 50 TCM-related databases. The main databases are listed as below.

1.2.1 Traditional Chinese Medical Literature Analysis and Retrieval System

The bibliographic system TCMLARS [25] has two versions. So far, the Chinese version contains over 600,000 TCM periodical articles, while the corresponding number reaches 92,000 in the English version of TCMLARS. The source material for the database is drawn from about 900 biomedical journals published in China since 1984. The main fields included in TCMLARS are similar to MEDLINE, such as title, author, journal title, publication year, and abstract. Besides, some fields specifically existing in TCM are also included, such as pharmacology of Chinese herbs, ingredients and dosage of formulae, drug compatibility, and acupuncture and Tuina points. TCMLARS is considered an important new asset in the literature review and meta-analysis of CHM by McCulloch et al. [26]. It also serves as a significant data resource for KDTCM, especially for the methods based on text mining.

1.2.2 Figures and Photographs of Traditional Chinese Drug Database

This database also has Chinese and English versions. The Chinese version contains over 11,000 records, while the English version contains 545 records. Each record represents a single herb, or mineral drug, or other natural medicine, and provides the cited information. The data is derived from the Chinese Materia Medica Dictionary, Thesaurus of Chinese Herbs, Chinese Medicinal Materials, Manual of Composition and Pharmacology of Common Traditional Chinese Medicine, and so on. The main contents of TCDBASE include drug name, original source of medical materials, collection and storage of medical materials, parts of the plant/animal for medicinal use, chemical composition, physical/chemical properties, processing methods, dosage form, pharmaceutical techniques, pharmacokinetics, toxicology, compatibility of medicines, efficacy, and adverse reactions and their treatment. Such contents of the TCDBASE make it an important collection for CHM research, as well as the KDTCM research related to CHM.

1.2.3 Database of Chinese Medical Formulae

This database contains more than 85,000 formulae. Each record represents a single prescription and provides the cited information. Data is derived from modern publications such as *Pharmacopoeia of the People's Republic of China* and *Chinese Medical Formula Dictionary*. The main contents of Database of Chinese Medical Formulae (DCMF) include formal name, efficacy, indication, usage, precaution, adverse reactions and its treatment, ingredients, modification of the prescription, dosage form and specifications, preparation, storage, compatibility of medicines, chemical composition, and toxicology. DCMF collects comprehensive clinical cases using combinatorial medicines used over thousands of years, thus is particularly worth an in-depth analysis by KDD technology, especially the approaches based on frequent itemset analysis and association rule mining.

1.2.4 Database of Chemical Composition from Chinese Herbal Medicine

This database contains over 4500 records. Each record represents a single chemical composition and provides the cited information. Data is derived from Active Compositions of Chinese Herbal Medicine, Pharmacology of Traditional Chinese Medicine, CHM, and so on. The main contents include formal name, chemical name, physical/chemical properties, molecular formula, chemical formula, origin, pharmacological action, efficacy, toxicity, adverse reactions, chemical category, and functional category. Although information of 3D chemical structure is not included in this database, Database of Chemical Composition from Chinese Herbal Medicine (DCCCHM) could still be used as an important data resource for CHM research and drug discovery from CHM.

1.2.5 Clinical Medicine Database

This database contains information on more than 3500 diseases. The source material for the database was drawn from authoritative reference books and teaching materials of Chinese and Western medicine. The main contents of Clinical Medicine Database (CLINMED) include name of diseases, disease classification code by Western medicine/TCM, disease name definition by Western medicine/TCM, Western medicine/TCM etiology, pathology, pathological physiopathology, pathogenesis, diagnosis, Western medicine therapy, treatment of TCM, CHM therapy, acupuncture and moxibustion, massage, and integrated therapy of Western medicine and TCM. CLINMED reflects the understanding and experience of treating diseases in modern China, involving both Western medicine treatment and TCM therapy. As an effective avenue to combining Western medicine and TCM, multilevel knowledge discovery could be carried out based on the integration of data in both fields, and CLINMED could contribute to this trend.

1.2.6 TCM Electronic Medical Record Database

The TCM Electronic Medical Record (EMR) contains the data of TCM clinical practice on inpatients and outpatients. Clinical daily practice plays a vital role in TCM research and theory refinement because, unlike modern medicine, almost no bench-side study is performed in TCM. The China government has initiated several important programs since 2002 to collect clinical TCM EMR data, and the clinical data warehouse TCM Electronic Medical Record Database (TCM-EMRD) was built for potential decision support applications and TCM knowledge discovery. Currently, TCM-EMRD contains more than 3500 EMRs of inpatients. The main contents of TCM-EMRD include TCM Diagnosis, CMF, and the conceptual description of symptoms. TCM-EMRD is especially ideal for clinical KDTCM studies because KDD was considered the major reason for establishing this database. Thus, special consideration is given to issues related to knowledge discovery, such as the quality of data and the standardization of terms.

Other data resources within the TCM-online Database System include a Database of Tibetan Medicines, Medical News Database, OTC Database, Database of State Essential Drugs of China, Database of Medical Research Awards in China, Database of Medical Products in China, TCM Pharmaceutical Industry Database, China Hospitals Database, and Database of Paired Drugs (DPD). The TCM-online Database System, as well as other digitized TCM data resources, serves as the available data sources in various KDTCM researches, which can be seen in the following review section.

1.3 Review of KDTCM Research

1.3.1 Knowledge Discovery for CMF Research

One distinguishing feature of TCM lies in its emphasis on the usage of combinatorial medicines, in the form of CMF. A CMF is composed of selected drugs and suitable doses based on syndrome differentiation for etiology and the composition of therapies in accordance with the principle of formulating a TCM prescription. Except for a very few single drugs in all the prescriptions used clinically, the great majority of them are compound drugs consisting of two or more drugs. The reason is that the potency of a single drug is usually limited, and some of them may produce certain side effects or even toxicity. But when several drugs are applied together, ensuring a full play of their advantages and inhibiting the disadvantages, they will display their superiority over a single drug in the treatment of diseases. In this way, TCM uses processed multicomponent natural products in various combinations and formulations. Due to the great diversity of candidate drugs for forming a compound prescription, hundreds of thousands of CMF have been accumulated over thousands of years.

To use combinatorial medicines properly, one key issue is to realize the combination rules of multiple drugs. Compared with Western biomedicine, countless TCM practices over thousands of years have accumulated numerous cases of combinatorial medicines in the form of formulae. These TCM formulae are valuable resources for research into drug compatibility. Besides, due to the traditional usage of combinatorial medicines, the analysis of combination rules of multiple drugs exhibits much more significance in TCM than in Western medicine. Therefore, CMF research attracts more data miners than other subjects.

A breakthrough in herbal combination rule research lies in the analysis of paired drugs, which usually means a relatively fixed combination of two CHMs. In a generalized definition of paired drugs, the number of medicines in paired drugs can be extended to three, four, and so on. According to TCM theory, such combinations can increase their medical effectiveness or reduce the toxicity and side effects of some drugs. A number of paired drugs have already been induced from practice and included in the DPD as a part of the TCM-online Database System recently. However, such kinds of valuable combinations could be revealed more deeply. When two drugs are frequently used in combination with each other in practice, they are more likely to be paired drugs. Therefore, frequent itemset analysis and association rule mining could be used to discover paired drugs [27–31]. As the most classical example of DM, association rule mining aims at searching for interesting relationships among items in a given dataset. It can serve as one proper KDD method to analyze combination patterns of multiple drugs in TCM. The first step in association rule mining, also the core issue in association rule mining, is to find frequent itemsets, which means each of these itemsets will occur at least as frequently as a predetermined percentage in the whole dataset. This method can be used to find frequent herb co-occurrences.

In 2002, the classical association rule method was used by Yao et al. [27] to study 106 formulae for treating diabetes. The results indicate that different experts have similar ideas and principles for treating this disease, which helps to reveal the scientific rule in the composition of the TCM formulae for diabetes. This work also demonstrates that KDD is a powerful tool for acquiring knowledge and expressing TCM knowledge in an understandable way. In 2003, Jiang et al. [28] used basic frequent itemset analysis and association rule mining to analyze 1355 formulae for the treatment of spleen–stomach diseases from the section of Prescriptions, Dictionary of Traditional Chinese Medicine. The results of KDTCM in Ref. [28] are found to correspond basically with the rules and characteristics of the compatibility of spleen–stomach formulae in TCM. Another association mining on spleen–stomach formulae was carried out by Li et al. [29] using an effective algorithm based on a novel data structure named indexed frequent pattern tree. However, the number of formulae used in the experiment was not reported in that literature. To provide a powerful assist for TCM experts and KDTCM participants, Li et al. [30] developed a CMF DM system named TCMiner in 2004. Several efficient algorithms were implemented for frequent itemset/association rule mining in TCMiner.

The largest frequent itemset analysis in the TCM field was carried out by He et al. [31] in 2004. In this study, the DCMF database in the TCM-online Database System, which contains 85,989 formulae collected over thousands of years, was used as the data source to discover potential paired drugs. To improve performance, an FP-growth algorithm was applied to find frequent herb co-occurrences in DCMF without candidate generation. Also, an important preprocessing step was carried out, that is, to remove *Radix Glycyrrhizae* from each formula. This is because *Radix Glycyrrhizae* is frequently used in all kinds of formulae (it can decrease or moderate medicinal side effects or toxicity and regulate actions of all other herbs in one formula). This research is also noteworthy in that all of the generated paired drugs are compared with DPD. The results of the discovered top 15 herb co-occurrences with highest frequency are presented in Table 1.1. The frequency and content of these herb co-occurrences are listed in the leftmost and rightmost columns, respectively. The middle column indicates whether these discovered herb co-occurrences exist in DPD. Those records with the field "Yes" substantiate the existing paired drugs statistically. More significantly, as is shown in Table 1.1, many frequent herb co-occurrences that do not exist in DPD could also be revealed, such as *Radix Angelicae Sinensis* and *Rhizoma Atractylodis Macrocephalae*. Such combinations are highly likely to be paired drugs, which are worthy of further analysis and verification by TCM experts.

Actually, the above process of knowledge discovery essentially simulates what TCM researchers did to generalize novel paired drugs in the past. The only distinction is that past discovery is based on human observations and experiences, while KDTCM participants utilize powerful computers and efficient algorithms instead to accomplish this task, which greatly quickens the disclosure of new paired drugs and promotes the development of compatibility rule research.

With the deeper understanding of frequent itemsets and association rules in the TCM background, researchers realize that it is not enough to discover patterns using conventional methods. Although itemsets with very high frequency always indicate significant discoveries in many disciplines, this is not the case in TCM formulae, just like the condition of *Radix Glycyrrhizae* mentioned above. To overcome this limitation, Zeng et al. [32] introduced a new support (maximum support) into association mining. Based on this new bi-support, as well as a bitmap matrix technique to improve efficiency, a new association mining algorithm named BM_DB_Apriori was proposed in Ref. [32]. The experimental results on 1060 formulae show that BM_DB_Apriori is much faster and more accurate than the baseline algorithm Apriori. Another improvement is the consideration of correlated patterns in TCM. In the basic approaches of frequent itemset/association mining,

Table 1.1 Top 15 Discovered Herb Co-occurrences with Highest Frequency

Frequency of Herb Co-occurrences	Exist in DPD	Discovered Frequent Herb Co-occurrences
5127	Yes	*Radix Ginseng, Rhizoma Atractylodis Macrocephalae*
4428	Yes	*Radix Angelicae Sinensis, Radix Ginseng*
4062	Yes	*Radix Angelicae Sinensis, Rhizoma Chuanxiong*
3523	No	*Radix Angelicae Sinensis, Rhizoma Atractylodis Macrocephalae*
3049	Yes	*Radix Ginseng, Radix Astragali*
2853	No	*Radix Ginseng, Poria*
2760	No	*Radix Angelicae Sinensis, Radix Sapshnikoviae*
2688	Yes	*Rhizoma Atractylodis Macrocephalae, Poria*
2678	Yes	*Radix Sapshnikoviae, Rhizoma Notopterygii*
2596	Yes	*Radix Angelicae Sinensis, Radix Astragali*
2522	No	*Radix Ginseng, White Poria*
2321	Yes	*Rhizoma Atractylodis Macrocephalae, Pericarpium Citri Reticulatae*
2210	Yes	*Radix Angelicae Sinensis, Radix Paooniae Alba*
2142	No	*Radix Ginseng, Radix Sapshnikoviae*
2127	No	*Radix Sapshnikoviae, Rhizoma Chuanxiong*

we aim to find itemset/rule with minimum support and confidence. However, for TCM formulae, the interesting itemsets/rules always have low support but high confidence. Besides, considering a paired drug A and B, if A and B frequently coexist in formulae, but the existence of A cannot increase the likelihood of the existence of B, it shows that B is very likely a general drug used to moderate side effects or regulate actions of all other herbs in many kinds of formulae (e.g., *Radix Glycyrrhizae*). As far as drug compatibility research is concerned, this is not what we are interested in. To remedy this situation, Zeng et al. [33] proposed a new method to discover bidirectional TCM association rules in 2005. A similar idea was presented and formalized by Zhou et al. [34] in 2006, and the concept of both associated and correlated pattern was presented in that paper. A new interesting measure of confidence was proposed for rationally evaluating the correlation relationships. This measure not only has proper bounds for effectively evaluating the correlation degree of patterns but also is suitable for mining long patterns. The experimental results on 4643 TCM formulae data demonstrate that mining considering both association and correlation is a valid approach to discover both associated and correlated patterns in TCM formulae.

One interesting extension of frequent itemset mining in CMFs appeared in 2005 [35]. Considering that most CHMs are natural products, the knowledge of botanical taxonomy could be introduced to analyze CMFs. Based on frequent itemset mining of paired drugs, as well as the mapping relation between CHM and the botanical family, Zhou et al. [35] performed experiments to discover family combination rules of CHM in formulae. Over 18,000 CMFs from TCMLARS were collected and five subtypes of these CMFs were chosen according to the efficacy such as promoting blood circulation for removing blood stasis, invigorating the spleen, and replenishing Qi. The discovery process was carried out for each subtype, and many family combination rules were revealed. For example, *Radix Astragali–Radix Angelicae Sinensis* is a typical CHM pair with the efficacy of invigorating the spleen and replenishing Qi, and Umbelliferae–Legaminosae builds the core of CMFs with this efficacy because the support of Umbelliferae–Legaminosae combination in CMFs with this efficacy is as high as 180%. Compared to CHM combination rules, family combination rules give a higher level of knowledge about the drug combinations, which would be of help in clinical CMF prescription practice and new drug development.

Both the frequent itemset and association rule mentioned above can be classified as a linear rule model. To explore and present the pattern of combinatorial medicines from different perspectives, researchers began to introduce other models. Considering that undirected graphical models are suitable for discovering causal relationships and associations between variables, Deng et al. [36] proposed a structural learning algorithm named Information-Based Local Optimization (IBLO) to analyze CHMs in 554 formulae for patients with apoplexy. These formulae were collected from books about historic prescriptions of Chinese medicine. A graph including the 40 most important herbs in these formulae was obtained by the IBLO algorithm, from which we can see which CHMs are frequently used together. This method is noteworthy for TCM in that it extends the pattern of combinatorial medicines research from a linear rule model to graph model.

The studies mentioned above share a common feature, that is, they generally analyze CMFs without the consideration of herb dosage. The reason behind it might be that in TCM literature, especially in historical literature, the dosage information of each herb in one formula is quite fuzzy. In some cases, the dosage is a range value (e.g., 20–30 g). In many other cases, the dosage information is even missing. New knowledge discovery approaches are needed to analyze formulae with such fuzzy dosage information in the historical literature.

One method of formula research with dosage is to study the dispensing ratio of different CHMs within one formula. Suppose a combination pattern of multiple CHMs is found to have therapeutic effects for a certain condition, the next issue is to determine the optimal dispensing ratio of these CHMs. This task can be performed with the help of knowledge discovery techniques. In 2003, Xiang [37] proposed a three-stage voting algorithm for mining the optimal herb dispensing ratio in formulae. This method was applied to obtain optimal *Radix Salviae Miltiorrhizae–Radix Notoginseng* dispensing ratio in the formulae to treat cardiopathy based on an animal model (dogs). Ten test groups including seven ratio groups (10/6, 10/3, 10/1, 1/1, 10/0, 0/10, and 1/10) and three comparison groups (groups of Western medicine, model of compound, and pseudo-surgery) were used in the experiments. The latter was designed to test the cardio-index of all samples. After extracting features about the therapeutic effects from time-series samples, the three-stage voting algorithm works as follows: the preliminary vote generates the features of each test group and cardio-index by each sample; the metaphase vote obtains the value of the therapeutic effect by each feature of the given group and index; the final vote mines the optimal dispensing ratio. An optimal dispensing ratio 10/6 was finally obtained in the experiment, showing the effectiveness of this method.

Confronted with the voluminous amount of TCM literature, historically accumulated and recently published, text mining is another group of knowledge discovery methods that can be used in KDTCM. Text mining is defined as the discovery of novel, previously unknown information, by automatically extracting information from different textual resources. For research of CMF, this technique was applied by KDTCM participants [35,38,39] to extract knowledge of herbs and formulae from TCM literature. Cao et al. [38] developed an ontology-based system for extracting knowledge of CHMs and CMFs from semi-structured text. In this work, ontologies of CHMs and CMFs were developed, consisting of a set of classes and their relations. In addition to offering a terminology for describing classes and their instances, the attributes of a class also play the role of knowledge placeholders for the class and its instances, and they are to be filled in during the knowledge extraction process. To perform knowledge acquisition, an Executable Knowledge Extraction Language (EKEL) was proposed to specify knowledge-extracting agents based on the guidelines of the frame-oriented ontologies, and a support machine was implemented to execute EKEL programs. The authors reported that the system successfully extracted knowledge of more than 2710 CHMs and 5900 CMFs. A limitation of this work lies in the requirement of semi-structured text as input. A more general method of text mining, based on the idea of bootstrapping, was utilized by Zhou and coworkers [39] in 2004. Given a small set of seed words and a large set of unlabeled data, this approach can automatically iterate to extract the objective patterns and new seeds from free texts. Based on this bootstrapping method, as well as other components (e.g., TCM literature DB and DM module), the authors developed a text mining system called MeDisco/3T (Medical Discover for Traditional Treatment Intelligence) [35]. In practice, MeDisco/3T was used in the family combination rule experiment [35] mentioned above, and the bootstrapping method was used as a significant step to extract CMF names, CHM components, and efficacy descriptions from TCMLARS with high precision.

1.3.2 Knowledge Discovery for CHM Research

Apart from CMF research, knowledge discovery techniques could also be used in CHM research. Existing work in this aspect can be further divided into two subfields: KDD for the research of CHM characteristics and KDD for the research of CHM chemical compositions.

1.3.2.1 KDD for the Research of CHM Characteristics

As one of the core contents in the TCM knowledge system, the completeness of the theory of CHM characteristics largely determines the accuracy and scientificity of forming a CMF and also the effectiveness and safety of this therapy. The therapeutic effects and side effects (toxicity) of the drugs are often two sides of the same coin. The toxicity of drugs must be thoroughly understood in order to ensure safety not only for orthodox medicine but also for CAM. Chen and Giu [40] applied computer-based statistical analysis to study the relevant factors of toxicity for 3906 kinds of commonly used CHM. Besides toxicity, seven other aspects of CHM data were collected, including nature and flavor, meridian tropism, processing method, botanical family, chemical compositions, the medicinal part, and pharmacological effects. The relations between these factors and CHM toxicity were analyzed by statistical analysis, yielding some meaningful results. Among 1119 CHMs with toxicity information recorded in the literature, only 3.3% was found to have high toxicity, providing evidence for the safety of most CHMs. However, 59.7% of the CHMs with a hot nature exhibit toxicity in varying degrees, showing that CHM toxicity is somehow related to the hot nature. It was also found that the botanical family of CHM is closely related to its toxicity. Among 108 kinds of CHMs belonging to Ranunculaceae, the proportion of CHMs with toxicity in varying degrees reaches as high as 42.6%. Moreover, 29.7% of Ranunculaceae CHMs exhibit high toxicity. Other families of CHMs with a relatively large proportion of toxicity include Araceae (14.2%) and Euphorbiaceae (11.3%). As for the factor of chemical compositions, the alkaloid is noteworthy in that 71.8% of CHMs with alkaloid ingredients exhibit toxicity. This number is by far the largest compared with other groups of chemical compositions in CHMs.

Another statistical analysis about CHM characteristics was performed by Yang [41] in 2005. In the study, 417 CHMs were collected from Chinese Pharmacopeia, from which 101 blood pressure-reducing CHMs were further selected. A comparison was made between these two groups of CHMs (101 groups and 417 groups) from the aspect of four natures, five flavors, and meridian tropism. The results show that most blood pressure-reducing CHMs are characterized by a pungent nature and bitter flavor, mainly acting on liver and gallbladder meridians. Such knowledge is beneficial to the discovery of effective phytochemical components and the research of pharmacological properties.

The relation between CHM efficacy and other characteristics is a hot topic in CHM research. The use of knowledge discovery approaches can help this research in part. Yao et al. [42] applied an artificial neural network (ANN) and decision tree to classify 54 deficiency-nourishing CHMs into four groups (four subtypes of deficiency-nourishing efficacy) in 2004. Two quantification methods of CHM characteristics were used in the experiments, including two-valued quantification and multivalued quantification. The classification results of KDD methods were compared with the records in TCM teaching materials as the evaluation. It was found that the classification accuracies of ANN equal the decision tree (98.11%) for multivalued quantification, but ANN (96.13%) outperforms the decision tree (87.10%) for two-valued quantification. The results show that different quantification methods of CHM characteristics have a certain influence on the prediction of efficacy, and ANN outperforms the decision tree in this task. However, the sample size in this experiment seems too small.

Besides statistical analysis and classification, clustering is another type of KDD approach used in research of CHM characteristics. Based on characteristic features, Zhou et al. [43] applied clustering to group 28 CHMs for relieving the exterior syndrome in 2004. By this clustering method, the authors obtained quite a few identical results to the CHM theory. Although the sample size was also small (28 CHMs), the methodology is still worth recognition. Another larger scale of clustering was carried out by He et al. [44] in 2004. In this KDTCM study, the TCDBASE in the TCM-online Database System was used as data source, which contains information of more than 11,000 CHMs. Different from the former experiment, this was an efficacy-based clustering, and the similarity of values in attribute efficacy was used to evaluate the closeness between CHMs. The clustering algorithm applied in the experiment was the agglomerative approach, also known as the bottom-up approach. As a hierarchical clustering method, this algorithm starts with each object forming a separate group. It successively merges the objects or groups close to one another into larger and larger clusters, until certain termination conditions. By this clustering, the large number of CHMs in the TCDBASE was clustered into different groups based on efficacy. Such kind of categorization is highly important for the research of CHM characteristics and the analysis of effective chemical components in CHM, because similar efficacies of different CHMs often indicate high closeness of characteristics, as well as the chemical ingredients contained therein. Furthermore, these clustering results can serve as valuable references for selection of substitute drugs in formula-forming and the design of novel CMFs.

1.3.2.2 KDD for the Research of CHM Chemical Compositions

The substances that exhibit therapeutic effects are the chemical compositions of CHM. However, because the medicinal powers of CHM come from the cooperative effects of multiple active ingredients, the conventional research methodology that targets only a single ingredient is not enough. One probing method for this problem is to analyze CHM by knowledge discovery techniques at the level of the chemical element. In 1998, Qi et al. [45] applied factor analysis and clustering to study the relationships between CHM characteristics and the contents of trace elements in CHM. Forty-two kinds of trace elements in 105 CHMs were determined and factor analysis was carried out. The results indicate that a 10-factor model can interpret the correlation of these trace elements. Also, the information of trace elements was used in a hierarchy clustering to classify 105 CHMs into different nature groups. Compared with traditional knowledge of the CHM nature, the trace-element-based clustering achieved an accuracy of 78.1%, showing that trace elements are related to the nature of CHMs to a certain degree. A follow-up study of trace elements was conducted by Qi et al. [46] in 2003. This time the information of trace elements was also used in a hierarchical clustering, but aimed at classifying 10 CHMs for treating exterior syndromes into different efficacy groups. The accuracy of classification in the experiment was 90%, partly revealing the correlation between efficacy and the amount of trace elements in CHMs.

The modern technique of fingerprinting enables us to take thousands of chemical compositions into account simultaneously, thus providing huge data resources for knowledge discovery. Due to its specificity, integrity, stability, and quantifiability, the fingerprint method has been used successfully in identification and quality evaluation of traditional Chinese medicinal materials. This is an application where knowledge discovery approaches can contribute a lot. Typical DM methods used in this field include primary component analysis (PCA), ANN, and fuzzy clustering [47]. The main drawback of these methods is that extracted features are needed from the original information to form the feature space. For fingerprint data, the dimensions of the feature space are up to 1000 after discretization, and the sample set is usually small. ANN easily suffers from overfitting on such a small train set, while the PCA method is more dependent on the statistical information after discretization. To solve this problem, a novel method based on nearest neighbor (NN) and a genetic algorithm was presented by Zhang et al. [48] in 2004. A new measure, named corresponding peak distance, was proposed to calculate the distance between samples directly for the NN classifier, and the genetic algorithm was used to optimize the parameters of the NN classifier. Experiments on High-Performance Liquid Chromatography (HPLC) data of *Radix Ginseng* indicate that this hybrid method can effectively identify the medicinal material from different harvest times or habitats.

Another KDTCM study with regard to CHM chemical compositions was conducted by Lu et al. [49] in 2005. The TCM chemical database [20] was used as the data source in this study, and association rule mining was carried out to analyze the relations between CHM efficacy, botanical family, bioactivity of chemical compositions, and pharmacology of CHM extracts. Some meaningful rules were obtained in the experiments. For example, a bidirectional association rule was found between the M-cholinoceptor blocking drug and Solanaceae. This rule indicates that a large proportion of plants in Solanaceae (21.37%) contain ingredients with M-cholinoceptor blocking activity and, meanwhile, a large proportion of M-cholinoceptor blocking drugs (15.69%) can be found from plants in Solanaceae. Another interesting rule was found between Ranunculaceae and an antihypertensive drug. Such kinds of rules found by knowledge discovery techniques are valuable for the drug discovery process.

1.3.3 Knowledge Discovery for Research of TCM Syndrome

The TCM Syndrome (Zheng) is one of the core issues in TCM; it is a holistic clinical disease concept reflecting the dynamic, functional, temporal, and spatial morbid status of the human body. Considering the systematic knowledge accumulated from thousands of years of TCM clinical practice, which is a valuable hypothesis for modern biomedicine research, it is of great importance to deepen the study of the TCM syndrome. One idea of syndrome research in this postgenomic era is to study the syndrome at the molecular level. As an effective avenue to combine Western medicine and TCM, knowledge discovery could be undertaken based on the integration of the literature and clinical data in both fields.

In a probing research connecting TCM with modern biomedicine, Wu et al. [39] proposed a text mining approach to identify the gene functional relationships from MEDLINE based on TCM knowledge stored in TCM literature. The TCM literature used in this study comes from TCMLARS, which contains more than 600,000 articles from 900 biomedical journals published in China since 1984. These materials were treated by a simple but efficient bootstrapping method to extract syndrome–disease relationships. Besides, the term co-occurrence was used to identify the relationships between disease and gene from MEDLINE. Then the authors obtained the syndrome–gene relationships by one-step inference, that is, to compute the genes and syndromes with the same relevant disease. The underlying hypothesis behind this research was that the relevant genes of the same syndrome would have some biological interactions.

The relationship of kidney YangXu syndrome and related genes was taken as an example by the authors because kidney YangXu syndrome is an important syndrome involving caducity, neural disease, and immunity. Moreover, through an experimental study of kidney YangXu syndrome at the molecular level, Shen [50] found that kidney YangXu syndrome was associated with the expression of CRF (C1q-related factor). This laboratory-confirmed relationship, as well as other novel syndrome–gene relationships, was found by the text mining method in Ref. [39].

The authors compiled about 2.5 million disease-relevant PubMed citations and 1,479,630 human gene-relevant PubMed citations in a local database drawn from online MEDLINE. Meanwhile, about 1100 syndrome–disease relationships were obtained from TCM literature published in 2002. Considering the bilingual issues, before searching MEDLINE, the Chinese disease name was translated partially automatically into a formal English disease name according to the TCM headings database, and a manual check by TCM terminological expert was conducted when no TCM headings for the disease existed. Filtering by total co-occurrence was more than 10, or it was discovered that the number of relevant diseases revealed more than 272 genes related to kidney YangXu syndrome, in which CRH (corticotropin-releasing hormone) was included. By querying the gene databases, it was found that CRF was an alias for CRH. Thus, it means the relationship between kidney YangXu syndrome and CRF had already been discovered by the method. Furthermore, suppose the alias relationship between CRF and CRH was not known, the authors showed that it was also possible to confirm the relationship between CRF and kidney YangXu syndrome by the next several steps. First, some important genes in kidney YangXu syndrome were selected, including CRP (C-reactive protein, pentraxin-related); CRH, IL10 (interleukin10); ACE (angiotensin I converting enzyme); PTH (parathyroid hormone); and MPO (myeloperoxidase). Second, the PubGene [51] was searched for a subset network using each of the above genes, and the corresponding subset networks were shown as in Figure 1.1. The third step was to analyze the extracted knowledge. Now suppose that we do not know CRF is a relevant gene of kidney YangXu syndrome. By analyzing the six subset networks in the left part and the CRF subset network in Figure 1.1, we may conclude that CRF is somewhat relevant to kidney YangXu syndrome because the subset networks, which reassembled with the gene nodes such as IL10, CRAT, and CRF/CRH/ACE/MPO/

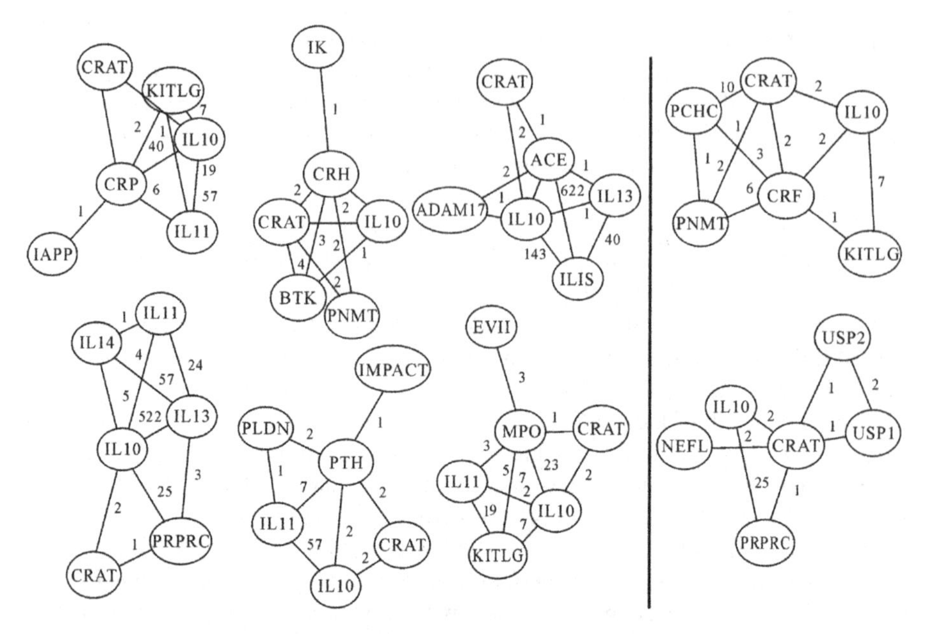

Figure 1.1 The six subset networks of each selected gene (left part of the vertical line). Gene CRAT, which is in all of the six subset networks, may be a novel relevant gene of kidney YangXu syndrome. The right part gives the subset network of the already known relevant gene CRF and the subset network of CRAT.

PTH, constitute possible functional gene communities that contribute to kidney YangXu syndrome. No existing literature reporting the relationship between CRF and kidney YangXu syndrome was used to generate the novel knowledge. It is exciting that this simple demonstration has shown that primary text mining results will largely decrease the workload in molecular level syndrome research.

This work proposes a tool for the TCM researchers to rapidly narrow their search for new and interesting genes of a specific syndrome. Meanwhile, the work gives specific functional information to the literature networks and divides large literature networks into functional communities (e.g., the community containing IL10, CRAT, and CRH/CRF, genes for kidney YangXu syndrome), which cannot be identified in the current PubGene. Moreover, through the syndrome perspective, the study provides an approach to subset selection or an explanation of a vast literature based on the gene network. The results demonstrate that the syndrome can give a novel top-down view of functional genomics research, and it could be a promising research field while connecting TCM with modern life science using text mining and other knowledge discovery methods.

1.3.4 Knowledge Discovery for TCM Clinical Diagnosis

Diagnosis is of crucial importance in any medical or healing system that works on the body. Unlike Western medicine, diagnostic methods in TCM include four basic methods (called Si Zhen in Chinese): inspection, listening and smelling, inquiry, and palpation. The case history, symptoms, and signs gained through those four diagnostic methods are analyzed and generalized to find the causes, nature, and interrelations of the disease and to provide evidence for further differentiation of syndromes. Although used in clinical practice for thousands of years, these diagnostic methods seem subjective and unreliable at the present time. Thus, researchers begin to apply modern techniques to improve the objectivity of these methods. Knowledge discovery is one of the techniques that could contribute greatly in this process.

Among the diagnostic methods mentioned above, the examination of the tongue is one of the most important approaches for obtaining significant evidence in diagnosing a patient's health condition. However, the clinical competence of tongue diagnosis was mostly determined by the experience and knowledge of the physicians adopting the tongue diagnosis and was easily influenced by environmental factors. Therefore, it is necessary to build an objective diagnostic standard for tongue diagnosis. In the last decades, researchers have been developing various methods and computer-based systems to solve this problem. A recent review of studies in pattern recognition of the tongue image was undertaken by Yue and Liu [52] in 2004. By analyzing these researches, we can find many cases where image-based knowledge discovery methods were used. A typical study involving knowledge discovery approaches was conducted by Pang et al. [53] in 2004. In this work, two kinds of quantitative features, chromatic and textural measures, were extracted from tongue images by digital image processing techniques, and Bayesian networks were employed subsequently to model the relationship between these quantitative features and diseases. Experiments on a group of 455 inpatients affected by 13 common diseases, as well as another 70 healthy volunteers, showed that the diagnosis accuracy based on the previously trained Bayesian networks is up to 75.8%. In another study in 2005, Ying et al. [54] proposed eight characteristic quantity variables of tongue manifestations, including color, shape, wetness, dryness, and so on. These features were treated by *k*-means clustering and ANN to predict the diagnostic result. An experiment was performed on 49 patients with cerebrovascular diseases and 39 healthy people, and the classification accuracies for *k*-means clustering and ANN based on the newly presented eight characteristics reached 87.5% and 92%, respectively.

Besides the diagnostic information obtained through tongue diagnosis, other measures, including all kinds of symptoms and signs, are also considered as potentially important features in TCM diagnosis. Unlike Western medicine in which intervention is performed only after a crisis arises, TCM is characterized by diagnosing diseases at earlier stages and subsequently preventing greater problems from occurring by adjusting the imbalance in the body in time. Therefore, symptoms that are not characteristic features for certain diseases in the biomedical background are also used in TCM diagnosis in many cases. The rationality of this diagnostic idea is partly supported by a recent study conducted by Lu et al. [55] in 2005. In this work, correlation between CD4, CD8 cell infiltration in gastric mucosa, *Helicobacter pylori* infection, and symptoms in 62 patients with chronic gastritis were analyzed by logistic regression analysis and *k*-means clustering. The results indicate that an assemblage of eight nondigestive-related symptoms (such as heavy feelings in head or body, thirst, and cool limbs with aversion to cold) could increase the predicted percentage of CD4 and CD8 cell infiltration in gastric mucosa, including lower CD4 infiltration by 12.5%, higher CD8 infiltration by 33.3%, and also non-*H. pylori* infection by 23.6%. This study also demonstrates that the subjective symptoms could play an important role in the diagnosis and treatment of diseases.

Apart from the studies of subjective symptoms in TCM diagnosis, some novel objective parameters for syndrome differentiation were also proposed by research, and usually these new parameters were further treated by knowledge discovery methods to establish the diagnoses. In a study conducted by Deng et al. [56] in 1996, the contents of 14 trace elements in hair of 163 cases of rheumatoid arthritis were used as the objective parameters in syndrome differentiation. Dynamic clustering was carried out based on these indexes, and the diagnostic results were compared with clinical diagnoses. Great consistency in this comparison was observed in the experiment, and the accuracy of the clustering-based method reached as high as 95.70%.

The above studies show the power of subjective symptoms and objective parameters in TCM diagnosis. However, this does not mean that all subjective symptoms and objective parameters should always be taken into account simultaneously. A recent study undertaken by Li et al. [57] in 2005 showed that the selection of symptoms could influence the accuracy of diagnosis in 209 patients with *H. pylori* infection. Thirty-five clinical presentations were observed in the experiment, including digestive symptoms (such as appetite, stomach ache, and nausea), general status (such as complexion and stool), psychological status (such as sleep and emotion), and pathogenic factors (such as smoking and alcohol). These parameters were analyzed by statistical methods and *k*-means clustering. In the experiment, the diagnostic accuracy based on 35 clinical presentations was 65.7%. It could be improved by 5.7% when only the assemblage of digestive symptoms was engaged or by 8.6% when the pathogenic factors, general status, and tongue observation were combined. The diagnostic accuracy could be decreased when only the general symptoms were engaged or when the pathogenic factors were accompanied by some common digestive symptoms. We can conclude from this study that we should be cautious about the selection of symptoms in diagnosis.

Research in the last decades showed that knowledge discovery methods combined with clinical experiments could greatly help to determine this selection. Existing KDD methods used in this task include factor analysis [58], clustering [58], rough set [59], Bayesian networks [60], and latent class analysis [61]. In an original research conducted by Zhang et al. [58] in 2005, the methods of factor analysis and clustering were combined in the diagnosis of TCM syndromes in 310 patients with posthepatitic cirrhosis. The information of the conventional four diagnoses was collected by the method of clinical epidemiological research, and these features were reduced and grouped by factor analysis. Based on these obtained common factors, clustering was carried out subsequently to form different groups of syndromes. The results indicate that the TCM syndromes in 287 of the 310 patients (92.58%) could be classified, and the 287 cases could be divided into seven categories of syndromes, showing the effectiveness of combining factor analysis and clustering. A rough set is another effective knowledge discovery method for attribute reduction. Thus is suitable for the selection of symptoms in TCM diagnosis. In 2001, Qin et al. [59] applied a rough set in the diagnosis of rheumatoid arthritis. The results show that the diagnostic accuracy of a rough set for rheumatoid arthritis is greatly higher than that of a fuzzy set. Because of its powerful capacity of handling uncertainty, a Bayesian network has become an attractive tool able to model the dependence and independence relationships among the variables in the domain with network. In a self-learning expert system constructed for TCM diagnosis in 2004, a novel hybrid learning algorithm GBPS based on Bayesian networks was proposed by Wang et al. [60]. This efficient algorithm was applied to discover the dependence and independence relationships among symptoms and essential symptoms (named key elements), and the results were represented by directed acyclic graphs. This explicit representation of Bayesian networks helps us gain an insight into TCM knowledge and explain the diagnosis and treatment of TCM experts. Besides Bayesian networks, this expert system also included other modules to extract knowledge from clinical data automatically, such as the module for mining frequent sets of key elements. This data-driven approach distinguished this system from the rule-based expert systems developed previously.

Another interesting knowledge discovery study in TCM diagnosis was conducted by Zhang and Yuan [61] in 2006. Unlike other research that focused on the selection of distinguishing symptoms, Zhang aimed at discovering the latent variables hidden in the TCM diagnosis. Considering that latent variable detection might lead to scientific discovery [62], Zhang [63] proposed hierarchical latent class (HLC) models to discover latent structures and applied HLC models to discover the latent variables in TCM diagnosis of kidney deficiency syndromes [61]. In this study, 2600 cases of the elderly, older than 60 years, were investigated to collect 67 symptoms related to kidney deficiency syndromes. The HLC model was subsequently learned by a search-based algorithm proposed in Ref. [64]. The resultant model was found to match the relevant TCM theories well, and diagnosis based on the model produced conclusions consistent with those of a group of experts. This indicates that latent structure models can help greatly in building an objective statistical foundation for TCM diagnosis.

1.4 Discussions and Future Directions

The previous sections aim at presenting a whole picture of the state of the art of KDTCM research. In this picture, a number of future research directions in KDTCM need to be highlighted. In this section, a discussion of these previous KDTCM studies is given, as well as a summary of the future directions in KDTCM.

Due to the preference and more understanding about combinatorial medicines, the TCM community provides a large number of real-world cases using multiple drugs as the form of CMF over thousands of years. With the digitalization of these data and the development of computer science, it is very natural for the research to introduce knowledge discovery techniques, especially the methods based on frequent itemset analysis and association rule mining, to discover compatibility rules of medicines in CMF data. This KDTCM research is inspiring, with plenty of room for improvement. The extension from conventional methods to correlated pattern mining and graph model learning adapts to the features of CMF analysis better. However, most of these studies still stay at the level of analyzing formulae without the consideration of CHM dosage. The research into the optimal dispensing ratio for CHMs is noteworthy, but new knowledge discovery approaches are still needed to analyze formulae with the fuzzy dosage information in the historical literature. Moreover, through experiments combined with knowledge discovery methods, the action target of different drugs in a formula could also be analyzed by observing how the dosage change in some drugs could influence the curability of the related syndrome and symptoms.

Besides dosage, current KDTCM research in CMF also suffers from the exclusion of some other important attributes of CMF, such as efficacy and indication. As a complex drug system, a CMF involves a large number of effective ingredients, which cooperate with each other and contribute to the final holistic therapeutic effects. Meanwhile, there are complicated relations among ingredients, efficacy, indication, and other elements within this system. A variety of important relationships between these attributes in CMF, such as the correlation between ingredients and indication, between efficacy and indication, and between ingredients and adverse reactions, are still waiting to be disclosed with the aid of knowledge discovery methods.

As for the research of CHM characteristics, the relations between medicinal effects and toxicity, as well as the connections among four natures, five flavors, meridian tropism, the botanical family, and other elements, are of vast importance to TCM community. Existing KDTCM studies demonstrate that it is very helpful in applying knowledge discovery approaches in this subfield. For example, the strong connections between some botanical families and their toxicities (e.g., Ranunculaceae) discovered by knowledge discovery methods [40] are highly beneficial to toxicological research. Processing is the preparation of crude medicinal materials according to TCM theory, which mainly serves as the function of reducing side effect/toxicity and promoting therapeutic effects. Currently, there are few KDTCM researches conducted with regard to processing methods. We believe that the future KDTCM studies in processing methods could help us to gain an insight into the traditional wisdom of TCM in the treatment of herb toxicity.

Compared with other subfields, KDTCM studies in chemical compositions of CHM are a bit preliminary. The analysis of trace elements by knowledge discovery methods seems interesting, but is not enough by far. Due to its specificity, integrity, stability, and quantifiability, the fingerprint technique has been successfully used in identification and quality evaluation of traditional Chinese medicinal materials, in which various knowledge discovery approaches are used to construct the classifiers. However, it is worth noting that the fingerprint technique also lays the foundation for DM of the spectra-effect relationship. Through analyzing the relativity of the variables generated by the fingerprint and the ones related to CHM characteristics and effects, it is promising to aid the discovery of active ingredients in CHM with known and unknown pharmacological actions. To derive new chemical drugs from CHM, the main approach is to extract potential effective ingredients from Chinese herbs as leading compounds, and then design new chemical drugs by structural modification on the basis of these leading compounds. However, during the discovery of lead compounds from natural products, the conventional random screening method suffers from low efficiency. For CHM, the ingredients with similar efficacy always have a resemblance in an active group. Thus, by Quantitative Structure-Activity Relationship (QSAR) analysis and related methods, knowledge discovery techniques could be utilized to seek promising active groups with regard to some therapeutic effects, which are beneficial to directional screening. However, there are few KDTCM researches currently on this aspect. With the increasing availability of high-quality three-dimensional (3D) chemical structure databases and the development of KDTCM techniques, we believe that knowledge discovery will contribute greatly to drug discovery in CHM.

Although the number of KDTCM studies in syndrome research is relatively small, it is very encouraging to see some studies connecting TCM with modern biomedicine in this subfield. In spite of the differences of methodology and technology, modern biomedicine and TCM share the same research subject, i.e., the disease phenomena of the human body. Thus, there are a lot of connections between biomedicine and TCM. For instance, on the one hand, the syndrome research could be carried out at the molecular level. On the other hand, the syndrome could give a novel top-down view of functional genomics research, which is particularly required in this postgenomic era. It is proved that text mining could contribute greatly in this multidisciplinary research. Based on the integration of the literature in both Western medicine and TCM, the syndrome–gene relationships, including some laboratory-confirmed relationships (e.g., CRF–kidney YangXu syndrome), are found discoverable by text mining. Moreover, through the syndrome perspective, current KDTCM study provides an approach to having a subset selection or an explanation of the vast literature-based gene network. As protein–protein interactions are central to most biological processes, the systematic identification of all protein interactions is considered as a key strategy for uncovering the inner workings of a cell. Thus, one aspect of future KDTCM work could focus on studying the protein–protein interactions using the knowledge of the TCM syndrome.

In spite of the interesting KDTCM work above in functional genomics research, it should also be noted that for TCM itself, currently the standard of the classification hierarchy for specific syndromes and even the definitions for some syndromes have not reached a consensus yet. Thus, the standardization and in-depth study of specific syndrome have become an urgent demand for development and research of TCM. This problem can also be partly solved by knowledge discovery approaches. For instance, information extraction could be carried out on the vast amount of

TCM literature, which helps to discover the most frequently used terms to describe syndromes in practice. All varieties of classification and clustering methods could be applied to syndrome data, which could provide useful guidance on how to form a classification hierarchy for a syndrome. More KDTCM research on the study of syndrome standardization can be expected in the future.

Apart from the above KDTCM directions mainly with regard to basic medical research, knowledge discovery techniques could also contribute to clinical TCM research. In practice, various kinds of DM methods have already been used in TCM clinical diagnosis, including factor analysis, clustering, ANN, rough set, Bayesian networks, and latent class analysis. Among these methods, Bayesian networks have several advantages. The main reasons are as follows. First, a Bayesian network can be used to learn causal relationships, and hence can be used to gain understanding about a problem domain and to predict the consequences of intervention, which is particularly suitable for analyzing the complicated features obtained by TCM diagnostic methods. Second, prior knowledge in TCM can be easily represented in Bayesian networks. Third, it is found [65] that diagnostic performance with Bayesian networks is often surprisingly insensitive to imprecision in the numerical probabilities. As an extension of conventional Bayesian networks, the HLC model can serve as a better tool to discover latent structures behind diagnostic observations in TCM. As reported in Ref. [61], currently the problem of the HLC model lies in the performance, which is only capable to deal with 35 variables in 67 candidates. In the future, this bottleneck can be expected to be removed with the development of related algorithms.

Another aspect of clinical research in TCM is the clinical evaluation. So far, there are hardly any knowledge discovery studies undertaken in this subfield. The reason probably lies in the unavailability of high-quality TCM clinical data sources in the past. However, the situation has been improving in recent years. The establishments of related real-world clinical databases, such as the TCM-EMRD mentioned in Section 1.2, will be of great help to promote the usage of KDTCM clinical evaluation in the future.

1.5 Conclusions

With a history that spans thousands of years, TCM provides Chinese people with effective health undertakings. As a medical system based on a holistic idea, which is totally different from orthodox medicine, TCM is a field worthy of in-depth analysis and research. What KDD is good at is searching the huge volume of data for meaningful patterns and knowledge, which is otherwise an almost impossible task to accomplish manually. Thus, KDD is a necessary technology to be applied in analyzing the great amounts of data in TCM. Based on an introduction to the current state of TCM data resources, this chapter provides an overview of knowledge discovery research in TCM in recent years. Major studies in four subfields of KDTCM are reviewed, including KDD for CMF research, KDD for CHM research,

KDD for TCM syndrome research, and KDD for TCM clinical diagnosis. The methods used in these studies are introduced, and some interesting results are highlighted. Finally, in a discussion section based on existing KDTCM studies, the current state and main problems of KDTCM work in each subfield are summarized, and the future directions for each subfield are also presented.

The journey of KDTCM in recent years is inspiring, resulting in considerably meaningful discoveries (such as the laboratory-confirmed relationship between CRF gene and kidney YangXu syndrome, the high proportion of toxic plants in Ranunculaceae, and the association between M-cholinoceptor blocking drug and Solanaceae). However, it should be noted that the KDTCM achievements are currently preliminary. For knowledge discovery, the TCM field is still nearly a piece of virgin soil with copious amounts of hidden gold. Here, the questions arise: where is the gold, what kind of gold is hidden, and how we can mine the gold? This chapter aims at answering these questions based on a review and analysis of existing KDTCM research, as well as a discussion of future directions. As indicated by Roddick et al. [11], mining over medical, health, or clinical data is arguably the most difficult domain for the KDD field. As a huge nonlinear complicated system, the human body always involves a large amount of mutual influence and dynamic balance among complex factors, which makes it extremely difficult to fully unravel the inner mystery of life phenomena. The advantage of TCM lies largely in the holistic thinking pattern and its preference for multicomponent therapy based on natural products. However, this also increases the complexity for KDTCM. Besides, currently the data of TCM still suffer from high individuality, ambiguity, and incompleteness. All this brings new problems and challenges when traditional KDD methods are applied to the TCM field. These problems are partly solved by the studies reviewed in this chapter. However, much more work needs to be done. Considering the ever-increasing volume of TCM data and the pressing demand to extract knowledge from these resources, we believe that the usage and development of KDTCM in the future will substantially contribute to the TCM community, as well as modern life science.

References

[1] Aderem A. Systems biology: its practice and challenges. Cell 2005;121(4):511–3.

[2] Eisenberg DM, Kessler RC, Foster C, et al. Unconventional medicine in the United States: prevalence, costs, and patterns of use. N Engl J Med 1993;328(4):246–52.

[3] Eisenberg DM, Davis RB, Ettner SL, et al. Trends in alternative medicine use in the United States 1097: results of a follow-up national survey. J Am Med Assoc 1998; 280(18):1569–75.

[4] Honda K, Jacobson JS. Use of complementary and alternative medicine among United States adults: the influences of personality, coping strategies, and social support. Prev Med 2005;40(1):46–53.

[5] Thomas KJ, Nicholl JP, Coleman P. Use and expenditure on complementary medicine in England: a population based survey. Complement Ther Med 2001;9:2–11.

[6] Yamashita H, Tsukayama H, Sugishita C. Popularity of complementary and alternative medicine in Japan: a telephone survey. Complement Ther Med 2002;10:84–93.
[7] Fayyad U, Piatetsky-Shapiro G, Smyth P. From data mining to knowledge discovery in databases. AI Mag 1996;17(3):37–54.
[8] Han J, Kamber M. Data mining: concepts and techniques. San Francisco, CA: Morgan Kaufmann Publishers; 2000.
[9] Bath PA. Data mining in health and medical information. Annu Rev Inform Sci Technol 2004;38(1):331–69.
[10] Altman RB. AI in medicine: the spectrum of challenges from managed care to molecular medicine. AI Mag 1999;20(3):67–77.
[11] Roddick JF, Fule P, Graco WJ. Exploratory medical knowledge discovery: experiences and issues. SIGKDD Explor 2003;5(1):94–9.
[12] Cios KJ, Moore GW. Uniqueness of medical data mining. Artif Intell Med 2002; 26(1–2):1–24.
[13] Zupan B, Lavrac N, Keravnou E. Data mining techniques and applications in medicine. Artif Intell Med 1999;16(1):1–2.
[14] Lavrac N. Selected techniques for data mining in medicine. Artif Intell Med 1999; 16(1):3–23.
[15] Cios KJ, editor. Medical data mining and knowledge discovery. Heidelberg: Springer-Verlag; 2000.
[16] Kononenko I. Machine learning for medical diagnosis: history, state of the art and perspective. Artif Intell Med 2001;23(1):89–109.
[17] Fan KW. Online research databases and journals of Chinese medicine. J Altern Complement Med 2004;10(6):1123–8.
[18] Liu Y, Sun Y. China traditional Chinese medicine (TCM) patent database. World Pat Inf 2004;26:91–6.
[19] <http://www.tradimed.com>; 2006 [accessed 17.05.06].
[20] Zhou JJ, Xie GG, Yan XJ. Traditional Chinese medicines: molecular structures, natural sources and applications. Burlington, VT: ASHGATE; 2003.
[21] Zhou XZ, Wu ZH, Lu W. TCMMDB: a distributed multi-database query system and its key technique implementation. In: Proceedings of IEEE SMC 2001, IEEE Computer Society; 2001. p. 1095–100.
[22] Chen HJ, Wu ZH, Huang C, et al. TCM-Grid: weaving a medical Grid for traditional Chinese medicine. In: Goos G, Hartmanis J, Van Leuwen J, editors. Proceedings of international conference on computational science 2003, lecture notes in computer science 2659. Berlin: Springer-Verlag; 2003. p. 1143–52.
[23] <http://www.cintcm.com>; 2006 [accessed 17.05.06].
[24] Zhou XZ, Wu ZH, Yin AN, et al. Ontology development for unified traditional Chinese medical language system. Artif Intell Med 2004;32(1):15–27.
[25] Fan WY. The traditional Chinese medical literature analysis and retrieval system (TCMLARS) and its application. Int J Spec Libr 2001;35(3):147–56.
[26] McCulloch M, Broffman M, Gao JM. Chinese herbal medicine and interferon in the treatment of chronic hepatitis B: a meta-analysis of randomized, controlled trials. Am J Public Health 2002;92(10):1619–27.
[27] Yao MC, Ai L, Yuan YM, et al. Analysis of the association rule in the composition of the TCM formulae for diabetes. J Beijing Univ Tradit Chin Med 2002;25(6):48–50 [in Chinese].
[28] Jiang YG, Li RS, Li L, et al. Experiment on data mining in compatibility law of spleen-stomach prescriptions in TCM. World Sci Technol Modern Tradit Chin Med 2003;5(2):33–7 [in Chinese].

[29] Li C, Tang CJ, Peng J, et al. NNF: an effective approach in medicine paring analysis of traditional Chinese medicine prescriptions. In: Zhou LZ, Ooi BC, Meng XF, editors. Proceedings of DASFAA 2005, lecture notes in computer science 3453. Berlin: Springer-Verlag; 2005. p. 576–81.

[30] Li C, Tang CJ, Peng J, et al. TCMiner: a high performance data mining system for multidimensional data analysis of traditional Chinese medicine prescriptions. In: Wang S, Yang DQ, Tanaka K, et al., editors. Proceedings of ER workshops 2004, lecture notes in computer science 3289. Berlin: Springer-Verlag; 2004. p. 246–57.

[31] He QF, Cui M, Wu ZH, et al. Compatibility knowledge discovery in Chinese medical formulae. Chin J Inf Tradit Chin Med 2004;11(7):655–8 [in Chinese].

[32] Zeng LM, Tang CJ, Yin XX, et al. Mining compatibility of traditional Chinese medicine based on bitmap matrix and bi-support. J Sichuan Univ (Nat Sci Ed) 2005;42(1): 57–62 [in Chinese].

[33] Zeng LM, Tang CJ, Yin XX, et al. Analysis of correlation based on bidirectional association rules. Comput Eng Des 2005;26(10):2585–8 [in Chinese].

[34] Zhou ZM, Wu ZH, Wang CS, et al. Mining both associated and correlated patterns. In: Alexandrov VN, Van Albada GD, Sloot PM, et al., editors. Proceedings of ICCS 2006, lecture notes in computer science 3994. Berlin: Springer-Verlag; 2006. p. 468–75.

[35] Zhou XZ, Liu BY, Wu ZH. Text mining for clinical Chinese herbal medical knowledge discovery. In: Hoffmann AG, Motoda H, Scheffer T, editors. Proceedings of DS 2005, lecture notes in computer science 3735. Berlin: Springer-Verlag; 2005. p. 395–7.

[36] Deng K, Liu DL, Gao S, et al. Structural learning of graphical models and its applications to traditional Chinese medicine. In: Wang LP, Jin YC, editors. Proceedings of FSKD 2005, lecture notes in computer science 3614. Berlin: Springer-Verlag; 2005. p. 362–7.

[37] Xiang ZG. A 3-stage voting algorithm for mining optimal ingredient pattern of traditional Chinese medicine. J Softw 2003;14(11):1882–90.

[38] Cao CG, Wang HT, Sui YF. Knowledge modeling and acquisition of traditional Chinese herbal drugs and formulae from text. Artif Intell Med 2004;32(1):3–13.

[39] Wu ZH, Zhou XZ, Liu BY, et al. Text mining for finding functional community of related genes using TCM knowledge. In: Boulicaut JF, Esposito F, Giannotti F, et al., editors. Proceedings of the 8th European conference on principles and practice of knowledge discovery in databases. Berlin: Springer-Verlag; 2004. p. 459–70.

[40] Chen XL, Gui XM. Quantitative analysis with multifactor for Chinese herbal medicine: relevant factors of toxicity. J Fujian Coll Tradit Chin Med 1995;5(1):27–30 [in Chinese].

[41] Yang GY. Analysis of the properties of 101 blood pressure reducing plants. J Henan Univ Chin Med 2005;20(118):22–7 [in Chinese].

[42] Yao MC, Zhang YL, Yuan YM, et al. Study on the prediction of the effect attribution of the deficiency-nourishing drugs based on the quantification of TCM drug properties. J Beijing Univ Tradit Chin Med 2004;27(4):7–9 [in Chinese].

[43] Zhou L, Tang XY, Fu C, et al. Fuzzy clustering analysis of Chinese herbs for relieving exterior syndrome. West China J Pharm Sci 2004;19(5):339–41 [in Chinese].

[44] He QF, Zhou XZ, Zhou ZM, et al. Efficacy-based clustering analysis of traditional Chinese medicinal herbs. Chin J Inf Tradit Chin Med 2004;11(7):561–2 [in Chinese].

[45] Qi JS, Xu HB, Zhou JY, et al. Factor analysis and cluster analysis of trace elements in some Chinese medicinal herbs. Chin J Anal Chem 1998;26(11):1309–14 [in Chinese].

[46] Qi JS, Xu HB, Zhou JY, et al. Studies on the amount of trace elements and efficacy in Chinese medicinal herbs for treating exterior syndromes. Comput Appl Chem 2003;20 (4):449–52 [in Chinese].

[47] Feng XS, Dong HY. Data mining in establishing fingerprint spectrum of Chinese traditional medicines. Prog Pharm Sci 2002;26(4):194–201 [in Chinese].

[48] Zhang LX, Zhao YN, Yang ZH, et al. Classifier for Chinese traditional medicine with high-dimensional and small sample-size data. In: Proceedings of WCICA 2004, IEEE computer society; 2004. p. 330–4.

[49] Lu AJ, Liu B, Liu HB, et al. Mining association rule in traditional Chinese medicine chemical database. Comput Appl Chem 2005;22(2):108–12 [in Chinese].

[50] Shen ZY. The continuation of kidney study. Shanghai: Shanghai Scientific and Technical Publishers; 1990. p. 3–31.

[51] Jenssen TK, Laegreid A, Komorowski J. A literature network of human genes for high-throughput analysis of gene expression. Nat Gen 2001;28(1):21–8.

[52] Yue XQ, Liu Q. Analysis of studies on pattern recognition of tongue image in traditional Chinese medicine by computer technology. J Chin Int Med 2004;2(5):326–9 [in Chinese].

[53] Pang B, Zhang D, Li N, et al. Computerized tongue diagnosis based on Bayesian networks. IEEE Trans Biomed Eng 2004;51(10):1803–10.

[54] Ying J, Li ZX, Li S, et al. Collection and analysis of characteristics of tongue manifestations in patients with cerebrovascular diseases. J Beijing Univ Tradit Chin Med 2005;28(4):62–6 [in Chinese].

[55] Lu AP, Zhang SS, Zha QL, et al. Correlation between CD4, CD8 cell infiltration in gastric mucosa *Helicobacter pylori* infection and symptoms in patients with chronic gastritis. World J Gastroenterol 2005;11(16):2486–90.

[56] Deng ZZ, He YT, Yu YM. Comparison between two diagnostic methods of computer's mathematic model and clinical diagnosis on TCM syndromes of rheumatoid arthritis. Chin J Int Tradit West Med 1996;16(12):727–9.

[57] Li S, Lu AP, Zhang L, et al. Anti-*Helicobacter pylori* immunoglobulin G (IgG) and IgA antibody responses and the value of clinical presentations in diagnosis of *H. pylori* infection in patients with precancerous lesions. World J Gastroenterol 2003;9(4): 755–8.

[58] Zhang Q, Zhang WT, Wei JJ, et al. Combined use of factor analysis and cluster analysis in classification of traditional Chinese medical syndromes in patients with posthepatitic cirrhosis. J Chin Int Med 2005;3(1):14–8 [in Chinese].

[59] Qin ZG, Mao ZY, Deng ZZ. The application of rough set in the Chinese medicine rheumatic arthritis diagnosis. Chin J Biomed Eng 2001;20(4):357–63 [in Chinese].

[60] Wang XW, Qu HB, Liu P, et al. A self-learning expert system for diagnosis in traditional Chinese medicine. Expert Syst Appl 2004;26(4):557–66.

[61] Zhang NL, Yuan SH. Latent structure models and diagnosis in traditional Chinese medicine. Technical report HKUST-CS04-12. Hong Kong: Department of Computer Science, The Hong Kong University of Science and Technology; 2006 [in Chinese].

[62] Zhang NL, Nielsen TD, Jensen FV. Latent variable discovery in classification models. Artif Intell Med 2004;30(3):283–99.

[63] Zhang NL. Hierarchical latent class model for cluster analysis. J Machine Learn Res 2004;5(6):697–723.

[64] Zhang NL, Kocka T. Efficient learning of hierarchical latent class models. In: Proceedings of ICTAI 2004, Los Alamitos; 2004. p. 585–93.

[65] Pradhan M, Henrion M, Provan GM, et al. The sensitivity of belief networks to imprecise probabilities: an experimental investigation. Artif Intell 1996;85(1–2):363–97.

2 Integrative Mining of Traditional Chinese Medicine Literature and MEDLINE for Functional Gene Networks[1]

2.1 Introduction

Functional genomics and proteomics have been the focus of post-genomic life science research. However, the reductionism approach is still the infrastructure of life science research because little holistic knowledge is valuable. Therefore, getting over the large-scale reductionism problem is a crucial issue for systems biology [1] research. Meanwhile, the last decade has been marked by unprecedented growth in both the production of biomedical data and the amount of published literature discussing it. As a distinguished biomedical data source, MEDLINE bibliographic database has accumulated over 12 million records since 2004. Automated literature mining offers a yet untapped opportunity to induce and integrate many fragments of information from multiple fields of expertise into a complete picture from which we can expose the interrelated relationships of various biomedical entities (e.g., genes and proteins). Many researches [2–9] have focused on the recognition of gene/protein name, gene–gene relations, and protein–protein interactions from biomedical literature (e.g., MEDLINE). Combining different kinds of biomedical data sources, such as expression, sequence, and literature data, is the mainstream of current integrative biomedical mining methods [10–13]. However, there is no such external knowledge as complete as traditional Chinese medicine (TCM) introduced in those researches. Herein, we integrate two distinct data sets, namely TCM literature and MEDLINE, to discover gene networks and gene functional knowledge.

TCM is an efficient traditional medical therapy (e.g., acupuncture and Chinese Medical Formula), which embodies holistic knowledge about human disease with thousands of years of clinical practice. Compared with the disease concept in modern biomedical science, TCM syndrome (simply referred to as "syndrome" in the following) is one of the core issues studied in TCM, which is a holistic clinical

[1] Zhou X, Liu B, Wu Z, Feng Y. Integrative mining of traditional Chinese medicine literature and MEDLINE for functional gene networks. Reprinted from Artif Intell Med 2007;41(2): 87–104. © 2007 Elsevier B.V., with permission from Elsevier.

Modern Computational Approaches To Traditional Chinese Medicine. DOI: http://dx.doi.org/10.1016/B978-0-12-398510-1.00002-9

disease concept reflecting the dynamic, functional, temporal, and spatial morbid state of the human body.

The traditional medical therapies based on syndrome differentiation propose a successful individualized medicine with convincible effects. Moreover, with the increasing research efforts put into the TCM clinical and medical issues, the relevant literature and fundamental medical data accumulated rapidly in the last century. Several significant TCM bibliographic databases have been built in TCM institutes and colleges since the 1990s. One main database is the TCM bibliographic database built by the Information Institute, China Academy of Chinese Medical Sciences, which contains over 500,000 records (up to 2005) from 900 biomedical journals published in China since 1984. The volume of the database increases by 40,000–50,000 records each year. There has been a literature retrieval system named TCMLARS [14] to let users take full-text search of the data. These high-quality literature storages can be good data sources for biomedical knowledge discovery research. Hence, integrating the two different kinds of huge data sources and using holistic knowledge such as the syndrome to understand or construct gene networks will give novel perspectives to systems biology research. This chapter aims at an integrative text-mining approach to identifying the gene functional relationships from TCM bibliographic literature and MEDLINE. The underlying hypothesis is that the related genes of the same syndrome will have some biological functional interactions. A simple and efficient bootstrapping method called bubble-bootstrapping is used to facilitate the extraction of syndrome–disease relationships from TCM literature. We obtain the gene nomenclature information from the HUGO Nomenclature Committee, which has 17,888 approved gene symbols. Using HUGO terminology and Medical Subject Headings (MeSH) disease headings, we generate the gene and disease MEDLINE literature index data. The term co-occurrence and pointwise mutual information-based relation weight computing method are used to identify the syndrome–disease relations in TCM literature and disease–gene relations in MEDLINE. The syndrome–gene relationships are identified by integrating the syndrome–disease relations and disease–gene relations with the connected disease terminology in both English and Chinese names. Then, the related gene networks of syndromes are searched, analyzed, and visualized by graph algorithms (e.g., graphviz2) and online analytical processing (OLAP) techniques. We have developed a prototype system named MeDisco/3S (Medical Discovery on diagnosiS Signature Synthesis) to implement the whole integrative text-mining process.

The rest of this chapter is organized as follows. Section 2.2 introduces the background of integrative mining of TCM literature and MEDLINE and the related key TCM concepts. In Section 2.3, we overview related research on biomedical literature mining. We introduce the name entity and relation extraction methods in Section 2.4. The components of the MeDisco/3S system are described in Section 2.5. In Section 2.6, we present the preliminary results and have a discussion of the gene network of Kidney Yang Deficiency (KYD) syndrome and functional analysis of genes from the syndrome perspective. Finally, we conclude the chapter in Section 2.7.

2.2 Connecting TCM Syndrome to Modern Biomedicine by Integrative Literature Mining

System biology [1] has been an active research field that seeks to integrate biological data to understand how biological systems function. By studying the relationships and interactions between various parts of a biological system (e.g., organelles, cells, physiological systems, and organisms), we hope that an understandable model of the whole system can be developed. However, it will be a critical problem if most of the current work remains just a new term for large-scale classical biological experimentation [15].

TCM is a complete system of medicine encompassing the entire range of human experience in China. Thousands of scientific studies that support traditional Chinese medical treatments are published yearly in journals around the world. TCM views the human body and the environment as a whole system. It embodies rich dialectical thoughts, such as Yin Yang Theory, Five Element Theory, and Zang-Fu Theory. Yin and Yang are complementary opposites used to describe how things function in relation to each other and to the universe. They are interdependent—one cannot exist without the other, and they have the ability to transform into each other [16]. The ideas of holism and Bian–Zheng–Lun–Zhi (synthesized treatment based on syndrome differentiation) are the fundamental infrastructure of TCM [17]. TCM studies the morbid state of human body by clinical practice and holistic qualitative cognitive process. Unlike the disease concept in modern medicine, syndrome reflects the comprehensive, dynamic, and functional morbid state of human body, which represents the TCM diagnosis result of symptoms and signs (TCM has developed a systematic approach to acquiring the symptoms and signs). In clinical practice, the relationship between disease and syndrome is a many-to-many one. Different patients with same disease could show different syndrome states, and one patient with the same disease in different stages also may have different syndromes. Moreover, one syndrome will occur in the patients with different diseases if the patients are in the same morbid state with the same manifestations. For example, KYD syndrome is a basic syndrome in TCM studies, which will refer to tens of hundreds of different diseases. It has been proved that diabetes has several syndromes, such as KYD syndrome, Qi–Yin Deficiency syndrome, Qi Deficiency Blood Stasis syndrome [18]. That is, different patients with diabetes may have different syndromes. Meanwhile, one patient with diabetes in different stages may display different syndromes too. Hence, the TCM prescription based on syndrome could be much different for different patients with the same disease or one patient with the same disease in different stages (this is one kind of individualized therapy in TCM). The traditional therapies (e.g., acupuncture and Chinese Medical Formula) based on syndrome are popular and effective in TCM clinical practice. Moreover, it is considered that the characteristics of the syndrome have much in common with that of genome and proteome such as polymorphism, individuality, and dynamics [19].We believe that the syndrome will provide much more functional and holistic knowledge about the human body, which will largely support the functional genomics and proteomics research as TCM has over thousands of years studied the syndrome state of the human body.

There are some studies such as that of Shen [20] on the molecular biology basis of the syndrome by experimental approaches. Shen has spent 50 years studying the KYD syndrome at the molecular level. He claimed that KYD syndrome is possibly located at the hypothalamus–pituitary–adrenal–thymus axis and he found that KYD syndrome is associated with the expression of CRF (corticotrophin-releasing factor) [21]. However, TCM syndrome is an abstraction and classification of the manifestations of patients. Although there have been experimental efforts at the molecular level fundamentals for the syndrome, it is still far from the target. The main obstacle is that there are still no good approaches to model syndrome, which is a human related dynamic and temporal qualitative organism concept, by animals. However, the integration of a biomedical literature knowledge source could be an effective and efficient method. Li et al. [22] have reported the work on Cold and Hot Zheng (syndrome in Chinese pronunciation) in the context of a neuro-endocrine-immune (NEI) network. They also evaluated the literature mining results by herbal treatment experiments on the rat model of collagen-induced arthritis. They found that the corresponding herbal treatments could affect the hub nodes of the Cold and Hot Zheng gene networks. Thus, they give a preliminary study on understanding Zheng in viewing the NEI network.

This paper suggests and proposes a synergistic approach for TCM and modern biomedical research by using integrative data mining techniques. We take the assumption that because modern biomedicine and TCM are both focusing on the study of disease phenomenon, we can regard the disease concept as the connecting point between modern biomedicine and TCM. The research on combination of syndrome and disease was systemically studied by Zhang Zhongjing, a famous ancient TCM clinical and theoretical figure in China, nearly 2000 years ago [23]. At present, the integration of disease diagnosis and syndrome differentiation has been an important mode for TCM clinical research [24]. There is a huge clinical bibliographic literature on the research of syndrome–disease relationships in TCM, which adhere to the holism principle of TCM theories. Meanwhile, the genetic pathology of disease is also intensively studied by modern biomedical research. Therefore, when analyzing the relationships between syndrome and gene/protein through disease concepts, we can generate some novel hypotheses, which are conceived neither in the TCM nor in the modern biomedical field. To have a pilot research on introducing TCM holistic knowledge to system biology, we focus on the connection of the syndrome and gene to get novel temporal and spatial gene interaction information, which cannot be easily acquired by large-scale genomics or proteomics techniques.

2.3 Related Work on Biomedical Literature Mining

The post-genomic topics have been the main issue in biomedical and life science researches. The human genomic sequence and MEDLINE propose the two most important shared knowledge sources, which will greatly contribute to the development and progress of modern biology research. Knowledge discovery from all kinds of huge biological data such as genomic sequence, proteomic sequence, and biomedical literature (e.g., MEDLINE, the annotations of Swiss-Prot, GenBank) has been the focus of bioinformatics research. Text mining from biomedical literature has been one of the most important methods for hypothesis-driven biomedical research.

Swanson [25] was one of the first researchers who used MEDLINE to find a novel scientific hypothesis. He proposed the concept of complementary literature probably with innovative knowledge [26] and developed a system named ARROWSMITH to help knowledge discovery in biomedical literature [27]. Gordon and Lindsay [28] and Lindsay and Gordon [29] applied information retrieval techniques to Swanson's early discoveries for literature-based discovery research. Weeber et al. [30] proposed the DAD-system, a concept-based natural language processing (NLP) system for PubMed citations, to discover knowledge of drugs and food. They claimed that the system could generate and test Swanson's novel hypotheses.

Most biomedical literature mining efforts to date have focused on biomedical entity classes, especially the identification of gene and protein names and the recognition of relationships among these entities. The recent research includes the recognition of gene/protein name [2,3,31–36], the identification of gene–gene relationships [4,9,37,38] and protein–protein relationships [5–8,39], and the identification of specific relationships between molecular entities, such as cellular localization of proteins [40], molecular binding relationships [41], and interactions between genes and drugs [42].

Utilizing text information to link molecular and phenotypic observations is also an important and promising research task. van Driel et al. [43] integrated different data sources to identify the candidate gene list underlying human genetic disorders and classified the phenotypes contained in Online Mendelian Inheritance in Man (OMIM) database to analyze the functional relations between genes and proteins [44]. Chun et al. [45] used a machine learning approach to enhance the performance of dictionary-based gene–disease relation identification. Masseroli et al. [46] developed a system, GFINDer, for the analysis of genetic-disorder-related genes by exploiting the information on genetic diseases and their clinical phenotypes present in textual form within OMIM database. Chen et al. [47] proposed a method to rank-order Alzheimer's disease (AD)-related proteins, based on an initial list of AD-related genes and public human protein interaction data. Perez-Iratxeta et al. [48,49] analyzed the relation of inherited disease and genes by MeSH terms of MEDLINE and RefSeq. While Freudenberg and Propping [50] acquired disease clusters according to the fuzzy similarity between phenotype information of disease, which was extracted from OMIM, then predicted the possible disease relevant genes based on the disease clusters. Antal et al. [51] used term co-occurrence from the literature to learn Bayesian networks as clinical models of ovarian tumors.

Due to the increment of publicly accessible biomedical data sources and the complexity of literature mining tasks, currently the integration of different biomedical data sources and usage of hybrid methods [10–13,43,45,47] have been the hot topics in biomedical literature mining.

Bunescu et al. [52] have a comparative study on the methods, such as relational learning, naive Bayes, support vector machine, in biomedical information extraction. de Bruijn and Martin [53] give an overview of a wide variety of applications and techniques for literature mining of biomedical texts. They regard literature mining as a modular process with four stages: document categorization, named entity tagging, fact extraction, and collection-wide analysis. They also discuss the related applications and methods from the four perspectives. Hirschman et al. [54] have a survey of biomedical literature data mining from the NLP perspective. They argue that there needs to be a challenging evaluation framework to boost the promising research. Hersh [55] also considers that better evaluation is essential for current biomedical text-mining systems. Yandell and Majoros [56] take a felicitous discussion of biomedical text mining and biological NLP as an emerging field that will be a great help to biology research.

Most recent work focuses on the extraction of terminology concepts or concept relationship knowledge, which has existed in the biomedical literature. Moreover, NLP techniques are preferred because the knowledge is conceived in the sentences. Such analysis is not only computationally prohibitive but also error prone when using NLP, and it is not applicable for large-scale literature mining. In this chapter, instead, we analyze large-scale literature using a simple method, which only considers the co-occurrence of terms. The most similar work is that of Jenssen et al. [9] and Wilkinson and Huberman [57]. They provide a simple approach to building a large-scale literature network of genes while only considering the gene co-occurrence as the view of gene interaction and using the dictionary-based term extraction method. Furthermore, Wilkinson and Huberman follow the method of Adamic et al. [58] to resolve some of the gene terminological ambiguities. However, we take advantage of TCM knowledge (e.g., syndrome and syndrome–disease relationship) to consider the related genes by a temporal or spatial holistic perspective. The method is based on the assumption that the genes related to the same syndrome would have some temporal or spatial connections because the syndrome reflects the holistic functional state of the morbid human body. Based on this assumption, currently our research has two aims. One is to find the relevant genes of the syndrome, and the other is to find the functional gene networks or gene functional information from the syndrome perspective. The bootstrapping technique is used to extract the disease names from TCM literature as no Chinese disease dictionary is available and the usage of disease terminology in the clinical literature is irregular. The experimental results show that bootstrapping is suitable for TCM terminology name extraction. Meanwhile, a dictionary-based term extraction method is used to build the gene- and disease-literature index from MEDLINE. The gene networks can be modeled as subgraphs as given in Refs. [9,57]. We believe that the syndrome as a core TCM clinical concept will give a novel approach to the discovery of gene functional relationships from the biomedical literature. This chapter proposes a prototype integrative literature mining system, MeDisco/3S, which incorporates bubble-bootstrapping, dictionary-based name entity extraction, relation weight computing, and OLAP methods, to implement the whole discovery process.

2.4 Name Entity and Relation Extraction Methods

This section describes the methods used in name entity extraction of diseases from TCM literature (in Chinese) and entity (e.g., gene, disease, and syndrome) relation weight computation. Because the capacity of syndrome terminology is limited and the usage in the literature is rather regular, we use a dictionary-based method to extract the syndrome names from TCM literature. Also, the same method is used for gene name identification in MEDLINE. To discover the syndrome–gene relation knowledge, we use the term co-occurrence and a relation weight computing method to extract the entity relationships among gene, disease, and syndrome. The following sections give an introduction to the bubble-bootstrapping and relation weight computing methods in detail.

2.4.1 Bubble-Bootstrapping Method

Bootstrapping is an iterative process to produce new seeds and patterns when provided with a small set of initial seeds. The initial seeds information gives the objective semantic type, which a bootstrapping technique should extract from the text. Current work was inspired by two papers, namely Yarowsky [59] and Blum and Mitchell [60]. Bootstrapping methods have been used to automatic text classification [61], database curation [62,63], and knowledge-based construction [64] from the World Wide Web. It should be noted that the problem addressed by bootstrapping is different from that of traditional information extraction, whose goal is to extract every instance from every document in which it occurs. On the contrary, the goal of bootstrapping is to extract just one instance of each tuple, each instance successfully extracted from at least one of the documents in which it occurs. In this section, we introduce a new bootstrapping method named bubble-bootstrapping.

As many terminological modifiers, which reflect the genre, state, or characteristic of a specific clinical disease, are used in clinical literature, and irregular clinical disease names are also popular, the clinical disease names are far more various than in a standard disease dictionary such as that of TCM headings terminology. Hence, the dictionary-based method is not applicable for disease name extraction in TCM literature. Moreover, Chinese is a language, which does not have space to indicate the word boundaries. Hence, word segmentation has been widely studied because it is a prerequisite for most NLP applications of Chinese texts [65]. However, accurate segmentation is challenging because of the segmentation ambiguity. This chapter gives a character-based bootstrapping method, which needs no word segmentation and shallow parsing process.

2.4.1.1 Pattern Definition

How to define, evaluate, and use pattern is the core issue of the bootstrapping technique. Because TCM literature is written in Chinese, the process of pattern will surely be different from that of bootstrapping used in an English text. This chapter takes a novel pattern definition and evaluation method and uses a search-match method without any shallow parsing processing for the extraction. We have designed one pattern definition named atomic pattern (ATP) to reflect the characteristics of TCM language. Compared with the previous research [66], we have an addition of a location-free feature character vector to the pattern. ATP is defined as

$$\text{ATP} = \left\langle \begin{array}{c} \text{leftPat, TermType, rightPat, RefCount,} \\ \text{FreqCount, Matched, VSM}_C \end{array} \right\rangle \tag{2.1}$$

in which

$$\text{VSM}_C = \{C_1 : M_{1,} C_2 : M_2, \ldots, C_n : M_n\}, M_1 > M_2 > \cdots > M_N \tag{2.2}$$

In Eqs. (2.1) and (2.2), Matched is a Boolean f to indicate the seed or pattern is used or not. leftPat represents the left Chinese string of the seed tuples, rightPat represents the right Chinese string of the seed tuples. Currently, we only extract two and three Chinese characters to form the left and right strings, respectively. TermType represents the semantic type of seed name, such as Chinese Medical Formula, disease, herb. RefCount and FreqCount represent the number of distinct seeds produced by pattern and the occurrence of pattern, respectively (we also define the corresponding RefCount and FreqCount of seed). VSM_C is the descending ordered character vector built by the sentence, which contains the relevant seed tuple. C_i is the Chinese character and M_i is the frequency of character C_i of the related ATP.

2.4.1.2 Bubble-Bootstrapping Algorithm

To keep the bootstrapping procedure robust when extracting the high-quality new seed tuples, we use a dynamic bubble-up evaluation method to ensure that the high-quality patterns and seed tuples will contribute to the iterative process. This bootstrapping method is called bubble-bootstrapping. We compute the RefCount and FreqCount of patterns and seed tuples dynamically and accumulatively, and let the patterns and seed tuples, which reach specific evaluation criterion, to attend the iterative procedure.

Currently, RefCount is computed before the next iteration and considered as the criterion to decide if the pattern (when RefCount is above a preassigned threshold) will attend the next step bootstrapping procedure. The global view of bubble-bootstrapping is depicted in Figure 2.1. We see that RefCount of patterns and seed tuples is dynamically increased during the bootstrapping circle. Therefore, the new patterns and seed tuples with enough RefCount value will emerge to boost the bootstrapping extraction process until the maximum iterative time is reached or all the RefCount value is below the defined threshold.

The two main components displayed in Figure 2.1 are the procedures of generating patterns and new seed tuples. Zhou [67] gives the detailed pseudo-code description of the two procedures. Table 2.1 shows the evaluation results of disease name extraction. The average precision of bubble-bootstrapping with ATP in disease

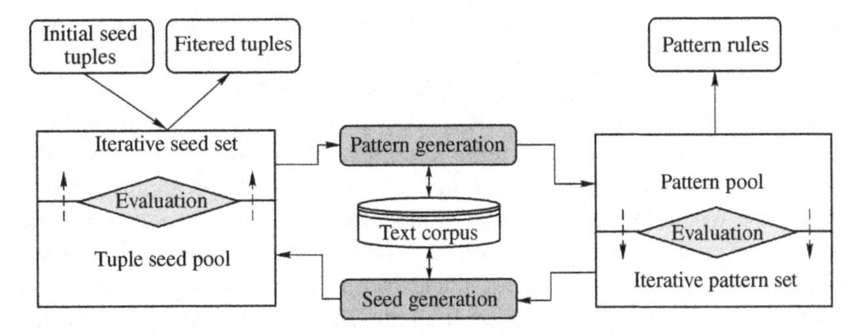

Figure 2.1 The components of bubble-bootstrapping. Given several initial seed tuples, bubble-bootstrapping approach uses these tuples to search the text corpus and generates corresponding patterns. After that, the selected patterns are used to search the text corpus to extract the new tuples. Then, the same procedure iterates using new selected seed tuples.

Table 2.1 Bootstrapping-Based TCM Terminological Name Recognition Results, the TCM Bibliographic Database of Years 1999, 2001, 2002, and 2003 has 38,937, 43,266, 44,315, and 16,151 Records, Respectively

Year	Term Type	Iterated P/S Threshold	Precision	Recall	Pattern Number	Seeds Number
2002	Disease	8/8	8/20	92.6%	3153	1018
2002	Disease	7/7	10/20	98.2%	3153	1097
2002	Disease	6/6	73/109	90.8%	5807	1753
2001	Disease	7/7	9/19	98.9%	2430	915
1999	Disease	7/7	4/19	99.4%	1459	853
2003	Disease	4/4	27/39	97.6%	1684	416

P/S is the abbreviation of pattern/seed.

name extraction is over 99.2%, and the average F_1 is about 64.4%. The results show that bubble-bootstrapping has high precision and needs very few seeds (not above 10 seeds) to boost an efficient unsupervised name extraction process. The extraction results of disease names help form a disease terminology database with 3445 records, which has the corresponding MeSH disease headings.

2.4.2 Relation Weight Computing

To handle the large data size, the term co-occurrence method is widely used in literature mining applications. The underlying hypothesis is that terms occurring in the same title or abstract would have some relationship. The strength of these relations can be evaluated by two popular methods. One is to make sure that the relation of two terms in the closer context is stronger. However, most of them would rely on the deep NLP of free text. It is still a bottleneck for large-scale processing applications. On the other hand, the term co-occurrence is a practical method for large-scale literature mining. Jenssen et al. [9] built the huge literature-based gene network, PubGene, with term co-occurrence, and Wilkinson and Huberman [57] also used the method. Furthermore, Jenssen et al. [9] let domain experts evaluate the gene relationships extracted by term co-occurrence and found that it has biological significance. In particular, if the term co-occurrence appears significantly more often than one would expect if it occurred independently at random, then it is likely that the gene relationships are correlated. Several statistical measures, such as pointwise mutual information [39], a model based on hyper-geometric distribution [68], were used to evaluate the relation strength. The related researches have shown that the term co-occurrence is an effective way to mine relations from collections of documents. However, the measure to evaluate the significant relations is important. We use the measure based on pointwise mutual information to compute the degree of syndrome–gene relations. The measure is defined as the following formula. Let $T_A = \{A_1, A_2, \ldots, A_i, \ldots, A_n\}$ and $T_B = \{B_1, B_2, \ldots, B_i, \ldots, B_m\}$ be two term sets, in which A_i and B_j are two observed terms. Hence, the weight of relation between A_i and B_j, $\mathrm{RW}(A_i, B_j)$, is defined as

$$\mathrm{RW}(A_i, B_j) = \frac{(F_{ij})^2}{\sum_{p=1}^{m} F_{ip} \sum_{q=1}^{n} F_{qj}} \tag{2.3}$$

where F_{ij} represents the co-occurrence frequency of A_i and B_j, while F_{ip} and F_{qj} represent the co-occurrence frequency of A_i and B_p ($B_p \in T_B$), and A_q ($A_q \in T_A$) and B_j, respectively. Furthermore, while considering the relations of syndrome and gene, we should introduce disease as midterm. Therefore, we define the relation weight of A_i and B_j, $\mathrm{RW}_{\mathrm{mid}}(A_i,B_j)$, if there are l midterms like C_k in midterm set $\mathrm{TC} = \{C_1, C_2, \ldots, C_k, \ldots, C_l\}$:

$$\mathrm{RW}_{\mathrm{mid}}(A_i, B_j) = l^2 \sum_{k=1}^{l} \mathrm{RW}(A_i, C_k)\mathrm{RW}(C_k, B_j) \tag{2.4}$$

where l represents the count of midterms between A_i and B_j. Given that the measure defined by Eqs. (2.3) and (2.4) mainly aims to rank the different relations and filter the relevant significant relations, this chapter uses the simple formula as much as possible. Because the syndrome–gene relationship is a many-to-many type, Eq. (2.4) tries to identify the significant syndrome–gene relations, which is more often than at random.

2.5 MeDisco/3S System

We have developed a prototype system named MeDisco/3S to discover and visualize the syndrome–gene relations and gene networks. The bubble-bootstrapping

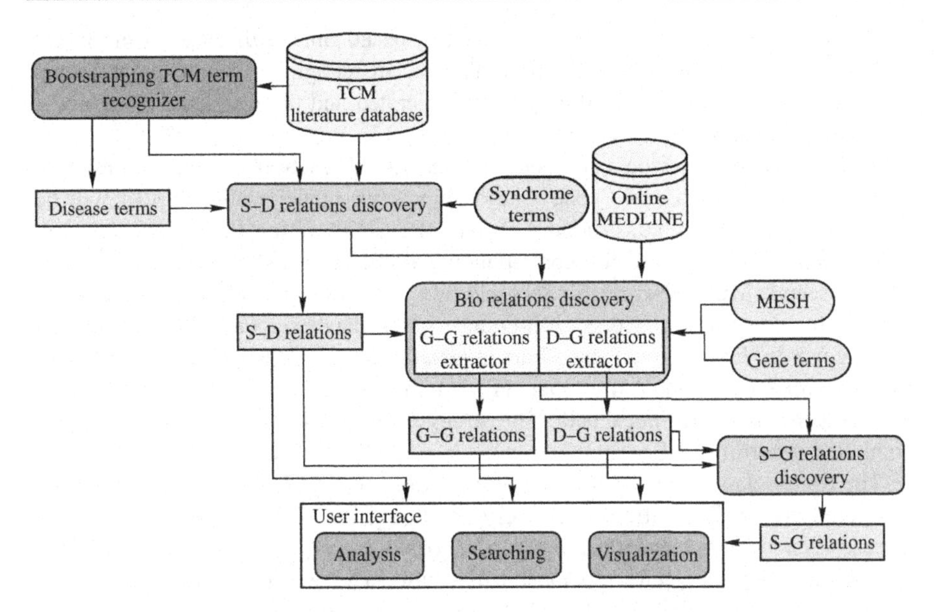

Figure 2.2 The general overview of MeDisco/3S system. The figure depicts the core functional modules and the related input data. Assisted by the terminological data (i.e., syndrome terms, MeSH, and gene terms), bootstrapping TCM term recognizer module, and the other three relation discovery modules (i.e., S–D relations discovery, bio-relations discovery, and S–G relations discovery modules), the relation data from TCM literature and MEDLINE was extracted. The relation data is stored in a local database, and then it can be analyzed, searched, and visualized by users.

method is used to extract the disease name entities in TCM literature. The relation weight computing method is used to evaluate the weight of syndrome–gene relationships, syndrome–disease relationships, and disease–gene relationships. Figure 2.2 shows the architecture of MeDisco/3S. It consists of five major modules: bootstrapping TCM term recognizer, S–D relations discovery, bio-relations discovery, S–G relations discovery, and user interface. We briefly introduce these core components in the following parts.

The bootstrapping TCM term recognizer module uses a bubble-bootstrapping algorithm to extract the disease name entities (in Chinese) from the title/abstract of TCM bibliographic records. Bubble-bootstrapping with ATP performs high-precision name detection tasks. However, some name entities will be too specific. The modification and mapping of these names to upper level terms should be conducted manually. Meanwhile, the alias names of some disease terms will be unified by experts based on the relevant terminological database (e.g., TCM headings and TCM clinical diagnosis/therapy terminology).

The S–D relations discovery module constructs syndrome-literature index data using syndrome terminology and a TCM literature database. The record of the index table has two fields: syndrome term identifier and literature identifier.

The disease-literature index data is also built based on the disease terminology, which is generated by a bootstrapping TCM term recognizer module. After that, the syndrome–disease relationships will be generated and the weight of syndrome–disease relationships will be computed using Eq. (2.3).

The bio-relations discovery module concerns extraction of the bio-relation knowledge, such as disease–gene relationship, gene–gene relationship (it will be used for gene network construction), from MEDLINE. Using Entrez utilities web service, this procedure use the corresponding MeSH terms (2004 version) of disease to search MEDLINE by MeSH field and download the related bibliographic records for constructing a disease-literature index database. Furthermore, we use the key words "human gene" and "human genetics" to query MEDLINE and download the related records. After that, HUGO-approved gene symbols are used to acquire gene-literature index data. The strength of disease–gene relations will be computed according to Eq. (2.3).

The S–G relations discovery module integrates the syndrome–disease relations generated by S–D relations discovery module from TCM literature, and the disease–gene relations generated by bio-relations discovery module from MEDLINE to form a syndrome–gene relationship database. The weight of syndrome–gene relations is calculated by using Eq. (2.4).

We have designed a data model of the relations data and stored all the generated relations data in a local database (using Oracle 10G). Thereafter, using visualization methods such as web OLAP, graphic libraries, we made all the data accessible by web interface. Currently, we use the business intelligence software, Business Objects XI platform, to develop the web interface with searching and analysis functionalities. Furthermore, we use graphviz to visualize the gene networks (using the gene-generation data generated by bio-relations discovery module).

Currently, we have implemented all the components of MeDisco/3S, which has Web OLAP functionalities to search and visualize the discovered relation knowledge among syndromes, diseases, and genes. The name entity extraction and relation weight computing procedure is a time-consuming task. However, the analysis and visualization procedure can be operated in real time. Figure 2.3 depicts the three kinds of relation data (syndrome–disease, disease–gene, and syndrome–gene relation data) displayed by web OLAP interface and describes how to integrate the syndrome–disease and disease–gene relation data to generate syndrome–gene relations.

2.6 Results

We have compiled about 2,500,000 disease-related PubMed citations and 1,479,630 human gene-related PubMed citations in a local database. All of the citations are drawn from online MEDLINE using NCBI Entrez utilities web service. We obtained over 13,000 disease–gene relations from MEDLINE. Meanwhile, we obtained 5600 syndrome–disease relationships from TCM literature (published

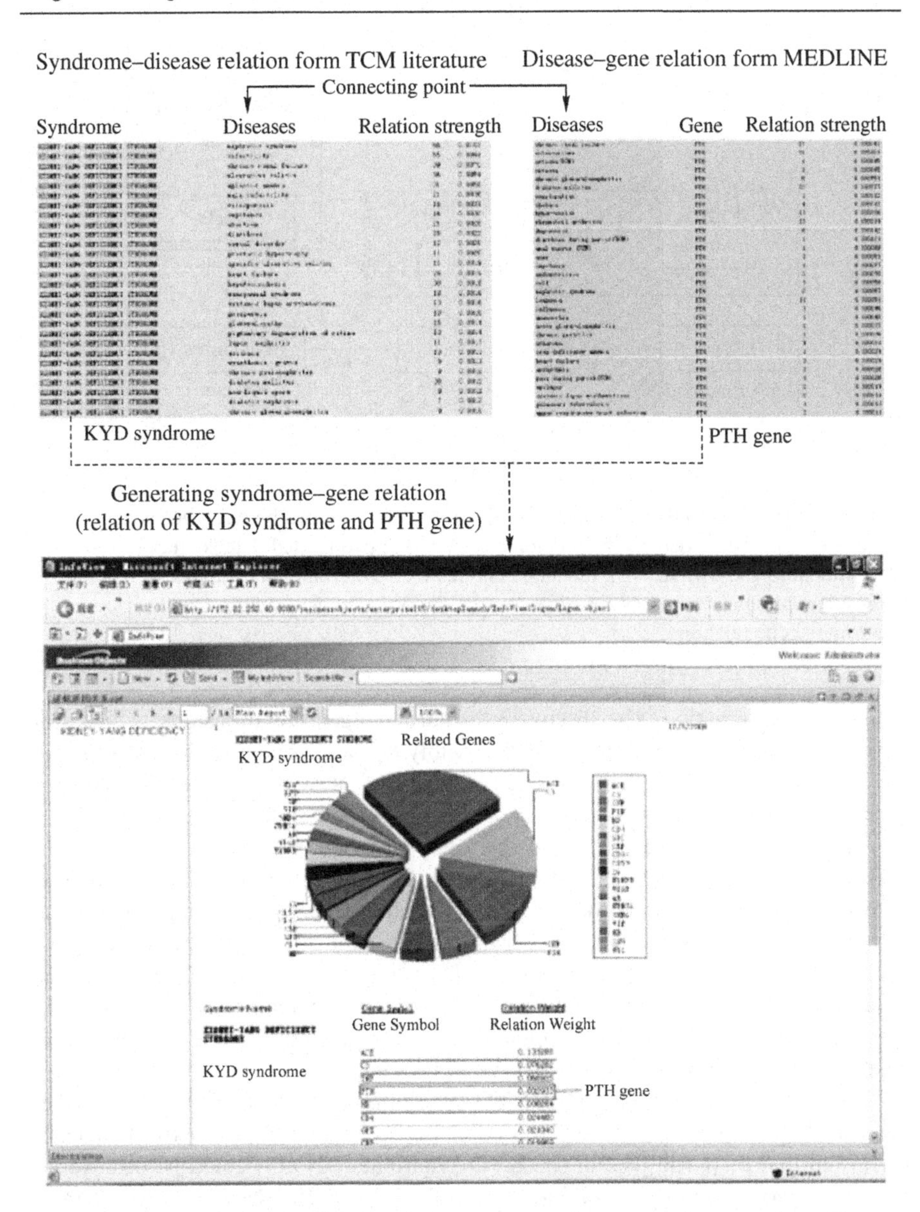

Figure 2.3 The logical view of the syndrome–gene relations generation procedure. The figure takes KYD syndrome as an example to demonstrate the procedure of syndrome–gene relation generated from syndrome–disease and disease–gene relations. The three types of relation data are displayed by MeDisco/3S OLAP.

from 1984 to 2005). Integrating the above two relations, we obtained about 200,000 syndrome–gene relations, whose relation strength is above zero. Figure 2.4 gives a picture of the concept associations supported by existing

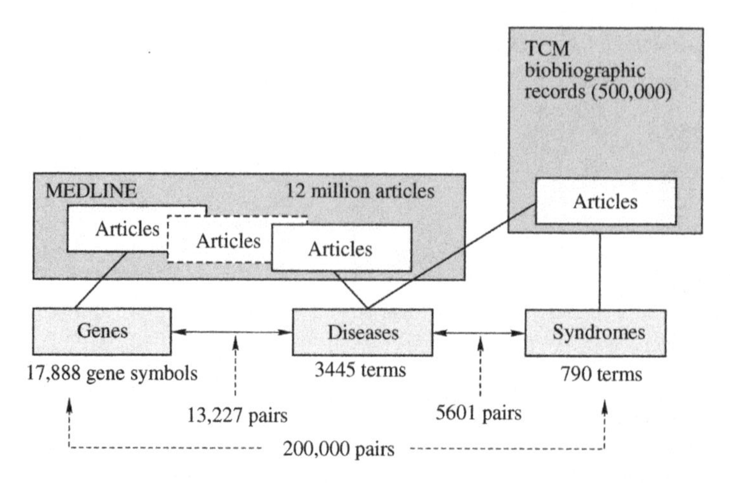

Figure 2.4 The data components used for identifying associations between genes and syndromes. The figure depicts the relational view of the two related data sources (i.e., TCM literature and MEDLINE), the name entities (i.e., syndrome, disease, and gene), and the result relations (i.e., syndrome–disease relation, disease–gene relation, and syndrome–gene relation).

biomedical information. The discovered syndrome–gene relationships, which still have no experimental evidence, would possibly be novel scientific hypotheses that are worthy of being further validated. Obviously, it will help boost syndrome research from a qualitative sense to the molecular quantitative level (Figure 2.5).

Furthermore, the related genes of the same syndrome possibly have some functional interactions, which are very different from the literature network that is purely built on gene term co-occurrence. Given the underlying assumption that the related genes of the same syndrome would have some functional interactions, we could discover the novel gene interaction knowledge even when the related genes have not co-occurred in MEDLINE yet. Furthermore, the appropriate subsets of the specific related genes propose an approach that is a selection of the huge literature gene network conceived in MEDLINE (e.g., PubGene).

In Ref. [66], we have taken the KYD syndrome as an example to demonstrate the novel gene knowledge discovery of the syndrome assisted by PubGene. It showed that the CRF gene of the KYD syndrome [21] also could be discovered by integrative mining of TCM literature and MEDLINE. Therefore, this chapter proposes an approach to obtaining novel syndrome-related genes compared with the experimental methods. Furthermore, the related genes of the syndrome can be a hypothesis and preliminary objective for further experimental efforts. The text-mining results will decrease the labor in molecular level research on the TCM syndrome.

With MeDisco/3S, we can generate more comprehensive results than the preliminary results of Ref. [66]. This chapter focuses on the gene function knowledge (e.g., gene networks and gene function predicting) discovered by MeDisco/3S. To be

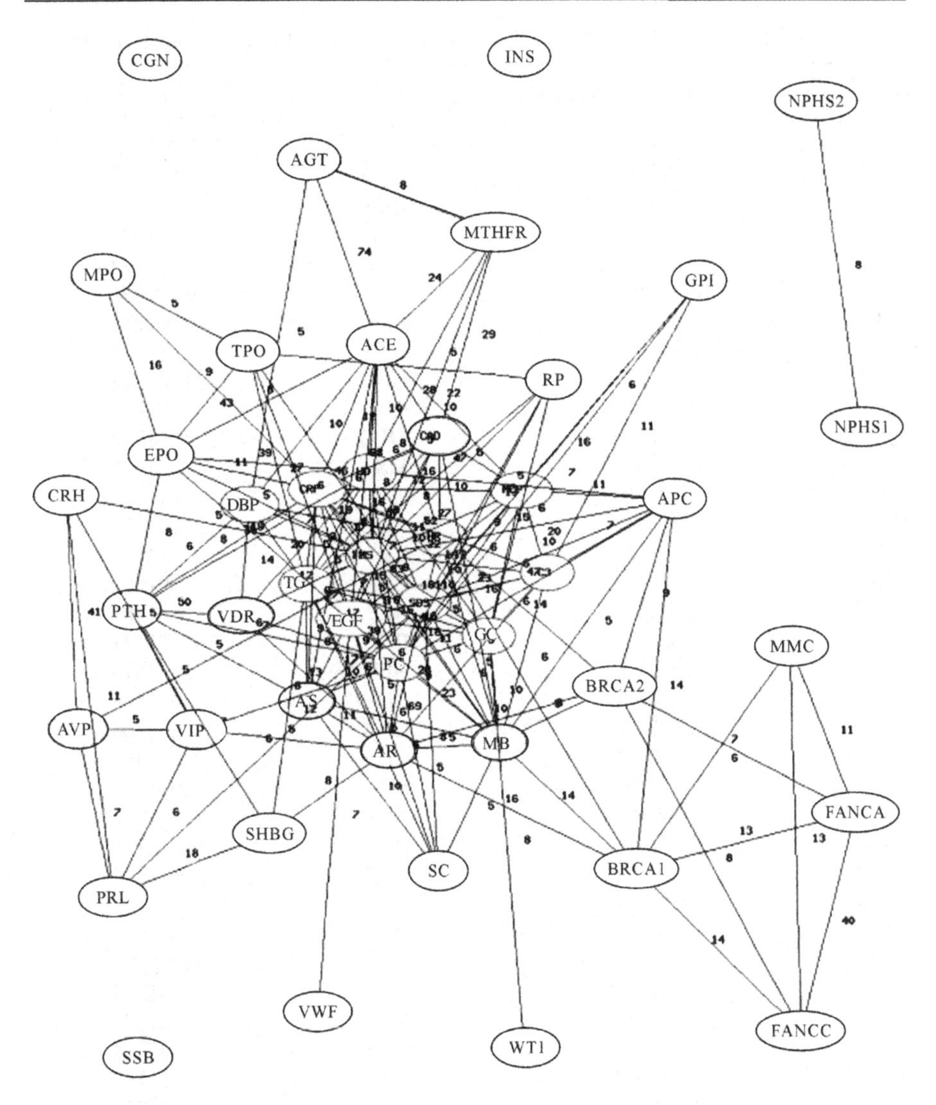

Figure 2.5 The gene network of KYD syndrome, in which the genes are ranked by relation weight as top 45. The number labeled on each edge represents the co-occurrence frequency of the related two genes.

more comparative, in this chapter, we still take the gene network of KYD syndrome as an example. Table 2.2 lists the latest results of 78 related genes of KYD syndrome with relation weight above 0.001, which is exported in Excel file format from the web interface of MeDisco/3S system. We see that the genes, such as ACE (angiotensin I converting enzyme (peptidyl-dipeptidase A)1), PTH (parathyroid hormone), TNF (tumor necrosis factor (TNF superfamily, member 2)), CD4 (CD4 molecule), GPI (glucose phosphate isomerase), CRP (C-reactive protein, pentraxin-related),

Table 2.2 The Top 78 Related Genes of KYD Syndrome Ranked by Relation Weight

Gene	Relation Weight	Gene	Relation Weight
ACE	0.133288	VWF	0.002808
C3	0.076282	CRH	0.002778
TNF	0.068915	FANCC	0.002662
PTH	0.032933	FANCA	0.002624
HD	0.030284	SSB	0.002414
CD4	0.024480	CGN	0.002376
GPI	0.021040	BRCA2	0.002281
CRP	0.019969	GC	0.002171
CD34	0.015618	APC	0.002053
CD59	0.014651	CD2AP	0.001851
TG	0.014466	MPO	0.001742
MTHFR	0.012544	MS	0.001675
VEGF	0.012178	CS	0.001652
AR	0.011185	TIMP3	0.001621
NPHS1	0.010377	C2	0.001594
SHBG	0.008111	CP	0.001591
VIP	0.007843	AFP	0.001487
MB	0.006963	IF	0.001471
INS	0.006684	NPHP1	0.001419
WT1	0.006295	CD19	0.001368
VDR	0.005741	GCA	0.001364
BRCA1	0.005689	CR1	0.001363
SDS	0.005599	SF	0.001284
PRL	0.005390	ADA	0.001280
CAD	0.004881	HFE	0.001279
MMC	0.004861	TRH	0.001274
NPHS2	0.004826	CD2	0.001267
EPO	0.004756	PTEN	0.001248
RP	0.004594	CD14	0.001220
AGT	0.004317	HP	0.001205
AVP	0.004299	SI	0.001180
CD68	0.004186	LPL	0.001164
HR	0.003560	CD28	0.001137
TPO	0.003505	PAP	0.001095
SC	0.003490	HGF	0.001087
DBP	0.003445	TAT	0.001074
MCP	0.003262	PP	0.001041
PC	0.003253	PKD1	0.001029
AS	0.003002	RET	0.001013

and VEGF (vascular endothelial growth factor), may have strong relationships to the KYD syndrome. Also, the gene CRH (corticotropin-releasing hormone, the alias name is CRF) is discovered with a relation weight 0.00278. The breast cancer genes such as BRCA1 (breast cancer 1, early onset), BRCA2 (breast cancer 2, early onset) are also discovered with rather high ranked order relation to the KYD syndrome. Because we use a dictionary-based method for gene name extraction, the listed gene names with two letters such as C3 (complement component 3), HD (huntingtin (Huntington disease)), TG (thyroglobulin), AR (androgen receptor), and MB (myoglobin) might be prone to false-positive errors.

This chapter provides a tool for TCM researchers to rapidly narrow their search for the syndrome-related genes. Meanwhile, we give the specific functional information to the literature networks and divide the large literature networks into functional communities (e.g., gene communities of KYD syndrome), which cannot be identified in the current PubGene. Moreover, through the syndrome perspective, we can easily have a subset selection or explanation of a giant literature-based gene network. Our approach will give two main supports for systems biology research. First, the approach can generate novel gene networks that have not been experimentally studied. Second, we can make it possible to predict the complete function of a gene by analyzing the gene relational distribution on syndromes. The following sections discuss the detailed results, respectively.

2.6.1 Functional Gene Networks

Compared with the work of Li et al. [22], this chapter focuses on the large-scale biomedical literature mining methods and all the human gene communities congregated by syndromes. We compute the syndrome–gene relationships in the context of all the syndromes and genes. Hence, the syndrome–gene relationship is more differential. The gene communities of a syndrome discovered by MeDisco/3S may constitute a functional gene network, which might represent the molecular basis of the syndrome. According to the genes of Table 2.2, MeDisco/3S uses graphviz graphic libraries to build the visualized gene network, as Figure 2.6 displays. We select the top 51 genes which exclude the genes namely TNF, CD4, CD59 (CD59 molecule, complement regulatory protein), CD68 (CD68 molecule), CD34 (CD34 molecule), CD2AP (CD2-associated protein) from all the following figures because they have wide literature co-occurrence relationships with the other genes and will largely decrease the presentation effect of the figures. Figure 2.6 only displays the edges between gene nodes if the two genes have above five-term co-occurrence in MEDLINE. From the gene network of KYD syndrome, we see that CRH has the edges to four genes namely AVP (arginine vasopressin (neurophysin II, antidiuretic hormone, diabetes insipidus, neurohypophyseal)), PRL (prolactin), VIP (vasoactive intestinal peptide), and MS (multiple sclerosis). It means that the genes CRH, AVP, PRL, VIP, and MS may have functional interactions. Even more, all the related genes of KYD syndrome may also have functional relations during some temporal period in individuals with the KYD syndrome relevant diseases. For example, although gene WT1 (Wilms tumor 1) has no edge to gene NPHS2 (nephrosis2, idiopathic, steroid-resistant (podocin)) in the current gene network, there could be a functional relationship (temporal or spatial) between WT1 and NPHS2 as the two genes are both significantly related to KYD syndrome.

KIDNEY-YANG DEFICIENCY SYNDROME	nephrotic syndrome	66	0.0103
KIDNEY-YANG DEFICIENCY SYNDROME	infertility	65	0.0091
KIDNEY-YANG DEFICIENCY SYNDROME	chronic renal failure	39	0.0070
KIDNEY-YANG DEFICIENCY SYNDROME	ulcerative colitis	36	0.0054
KIDNEY-YANG DEFICIENCY SYNDROME	aplastic anenia	31	0.0052
KIDNEY-YANG DEFICIENCY SYNDROME	male infertility	21	0.0036
KIDNEY-YANG DEFICIENCY SYNDROME	ostooporosis	16	0.0031
KIDNEY-YANG DEFICIENCY SYNDROME	importance	16	0.0030
KIDNEY-YANG DEFICIENCY SYNDROME	abortion	21	0.0028
KIDNEY-YANG DEFICIENCY SYNDROME	diarrhoea	26	0.0022
KIDNEY-YANG DEFICIENCY SYNDROME	sexual disorder	12	0.0020
KIDNEY-YANG DEFICIENCY SYNDROME	prostatic hypertrophy	11	0.0020
KIDNEY-YANG DEFICIENCY SYNDROME	spacific ulcarative colitix	11	0.0019
KIDNEY-YANG DEFICIENCY SYNDROME	heart failure	28	0.0019
KIDNEY-YANG DEFICIENCY SYNDROME	hepatocirrhoxix	30	0.0018
KIDNEY-YANG DEFICIENCY SYNDROME	menopauxal syndrone	18	0.0016
KIDNEY-YANG DEFICIENCY SYNDROME	systemic lupus erythematosus	13	0.0016
KIDNEY-YANG DEFICIENCY SYNDROME	prospernia	10	0.0016
KIDNEY-YANG DEFICIENCY SYNDROME	glonerulopathy	15	0.0014
KIDNEY-YANG DEFICIENCY SYNDROME	pigentary degeneration of ret	13	0.0014
KIDNEY-YANG DEFICIENCY SYNDROME	lupus nephritis	11	0.0013
KIDNEY-YANG DEFICIENCY SYNDROME	acidosis	10	0.0013
KIDNEY-YANG DEFICIENCY SYNDROME	myasthenia gravis	9	0.0013
KIDNEY-YANG DEFICIENCY SYNDROME	chronic pyelonephritis	9	0.0012
KIDNEY-YANG DEFICIENCY SYNDROME	diabetes nellitus	38	0.0012
KIDNEY-YANG DEFICIENCY SYNDROME	non-liquid sperm	9	0.0012
KIDNEY-YANG DEFICIENCY SYNDROME	diabetic nephrosis	7	0.0012
KIDNEY-YANG DEFICIENCY SYNDROME	chronic glonarul onephritis	9	0.0010
KIDNEY-YANG DEFICIENCY SYNDROME	sick sinus syndrone	8	0.0010
KIDNEY-YANG DEFICIENCY SYNDROME	spernatorrhas	11	0.0010
KIDNEY-YANG DEFICIENCY SYNDROME	habitual abrotion	5	0.0010
KIDNEY-YANG DEFICIENCY SYNDROME	amenorrhra	13	0.0010
KIDNEY-YANG DEFICIENCY SYNDROME	hypotthyroidisn	4	0.0010
KIDNEY-YANG DEFICIENCY SYNDROME	hyperplasia of endometrium	1	0.0009
KIDNEY-YANG DEFICIENCY SYNDROME	vasovagal syncope	1	0.0009
KIDNEY-YANG DEFICIENCY SYNDROME	hypertension	34	0.0009
KIDNEY-YANG DEFICIENCY SYNDROME	hyperprolactinemia	4	0.0009
KIDNEY-YANG DEFICIENCY SYNDROME	non-ejaculation	6	0.0008
KIDNEY-YANG DEFICIENCY SYNDROME	Alrheimer's diseaxe	5	0.0008
KIDNEY-YANG DEFICIENCY SYNDROME	dyxfunctional uterine bleedin	5	0.0008
KIDNEY-YANG DEFICIENCY SYNDROME	tinnitus	18	0.0008

Figure 2.6 Part of relevant diseases of KYD syndrome that are ranked in descending order by relation weight.

Furthermore, the significant novel gene relations, which have not been reported in the literature, are the promising hypotheses for further experimental research.

The syndrome is a qualitative diagnosis with different levels (e.g., mild KYD syndrome and severe KYD syndrome). There might be different inducements or pathologies to one common syndrome. Therefore, we can make the assumption that not all the relevant genes should be in the deviant state when the patients have one syndrome. Because the syndrome–disease relationship is a many-to-many type in TCM clinical practice, we can image that the molecular basis of syndrome and disease is a mode of two different functional gene/protein sets with some common elements. Hence, we can divide the whole gene network of one syndrome to different subset networks according to the relevant diseases of the syndrome. The subset networks corresponding to the relevant diseases could represent the different molecular inducements or pathologies for the syndrome. For example, we can easily get the relevant diseases of KYD syndrome by MeDisco/3S web interface (Figure 2.6). Here we see that the diseases such as chronic glomerulonephritis, chronic renal failure, aplastic anemia, heart failure, systemic lupus erythematosus, amenorrhea (one of the relevant diseases of gene CRH), and menopausal syndrome are related to KYD syndrome. As Figure 2.7 depicts, we get seven subset gene networks of KYD syndrome according to the seven different diseases. There are several common genes such as ACE, PTH, and CRP in the gene networks of chronic glomerulonephritis, chronic renal failure, and heart failure. Also, the diseases such as heart

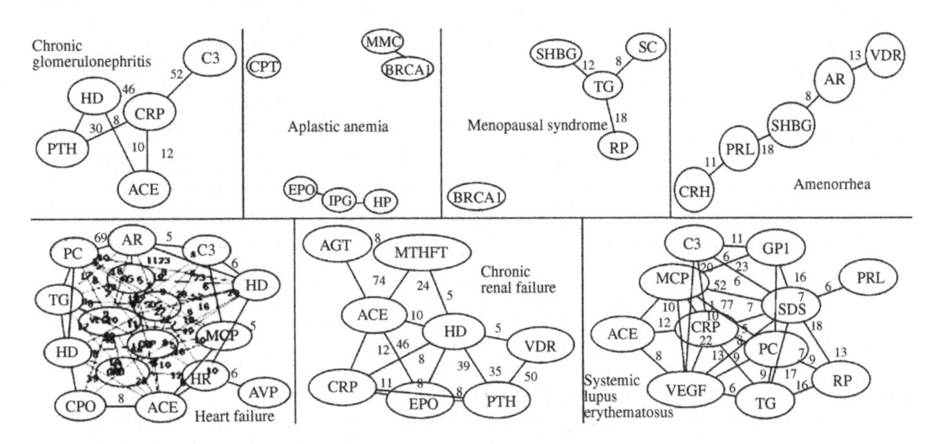

Figure 2.7 The gene networks of the six relevant diseases of KYD syndrome, in which all the genes are relevant to KYD syndrome.

failure and systemic lupus erythematosus have ACE, VEGF, CRP, TG, MCP (membrane cofactor protein (CD46, trophoblast-lymphocyte cross-reactive antigen)), SDS (serine dehydratase), and PC (pyruvate carboxylase) as common genes. However, most of the genes are different. This is partially due to the current gene co-occurrence in MEDLINE, i.e., the current research actually might have not paid sufficient attention to some relevant genes in these diseases. Whereas, according to TCM theories, the difference is inevitable and more significant because one syndrome could be induced by different pathologies. These subset gene networks provide operable and straightforward objectives for experimental approaches on the patients with KYD syndrome and the relevant diseases. For example, the gene network of chronic renal failure in Figure 2.7 represents the possible gene network of chronic renal failure patients with KYD syndrome. Because the syndrome reflects the morbid state (abstraction and classification based on manifestations of patient) of live individuals, this kind of gene network grasps more individualized, dynamic, and concrete morbid molecule information than disease-oriented gene networks using traditional, biological, experimental approaches. Currently, the syndrome–gene relations data involve about 400 syndromes. Like KYD syndrome, all the other syndromes can generate the corresponding functional gene networks with biological significance.

2.6.2 Functional Analysis of Genes from Syndrome Perspective

It is evident that most genes and their corresponding products have multiple functions. Therefore, discovering the complete function of a gene in human cells will be a permanent activity of biology research. As discussed in Section 2.2, the syndrome is the main clinical diagnosis of TCM. It has a big difference from modern disease diagnosis. The syndrome reflects the individualized and dynamic state of

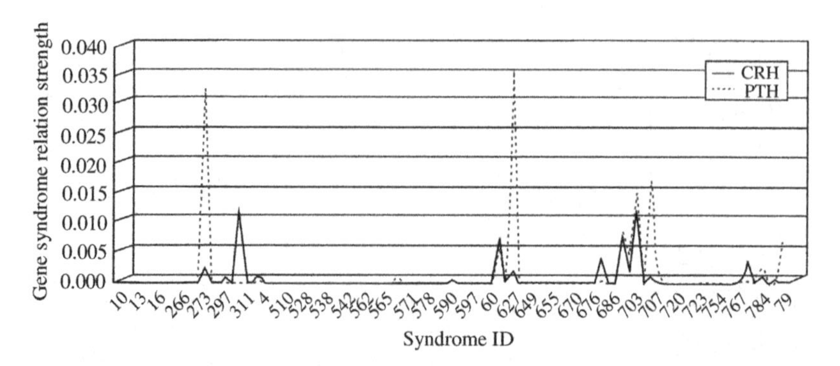

Figure 2.8 Gene relation strength distributions on syndromes. The figure displays the comparative distributions of the two genes CRH and PTH. It should be noted that the listed data is of several pages and the figure is the first page.

patients. The total number of basic syndromes in TCM is limited. In essence, the basic syndromes give a general classification of human morbid states. Hence, it will be helpful to gene function analysis while looking at the gene function from the syndrome perspective.

After the syndrome–gene relations have been constructed, we can analyze and predict the functions of a gene from the syndrome perspective. One simple way is to use the syndrome–gene relation strength distribution information. We assume that genes with the same function or in the same biological pathway may show similar relational distributions on syndromes. We have used OLAP tools to display the gene relational distributions on syndromes such as those in Figures 2.8–2.10. From the curve graphic figure and the listed data, a researcher with TCM knowledge can easily predict the gene function or compare the functions of two or three more different genes. The underlying assumption can be that while the relational distributions of two genes have similar curves, they may have similar functions and vice versa. This will provide the chance and hypothesis for novel gene function experimental research.

For example, Figure 2.10 displays the similarity between BRCA1 and BRCA1 genes. It shows that although most corresponding weights of BRCA2 on syndromes are less than the weights of BRCA1, they have a similar curve on the different syndromes. Hence, we can consider that the genes BRCA1 and BRCA2 may have similar functions. Furthermore, it shows that BRCA1 and BRCA2 genes are mainly correlated to the syndromes, such as KYD syndrome, Yin Deficiency syndrome, Kidney-Yin Deficiency syndrome, and Spleen–Kidney Deficiency syndrome. Hence, according to TCM syndrome knowledge, we can predict that the functions of BRCA1 and BRCA2 may be relevant to the reproductive system, developing system, endocrine system, digestive system, etc. However, it shows that the functions of CRH and PTH are significantly different (Figure 2.8), and so are the functions of EPO and AVP, and EPO and PRL (Figure 2.9). The functions of these genes can also be analyzed using TCM syndrome knowledge.

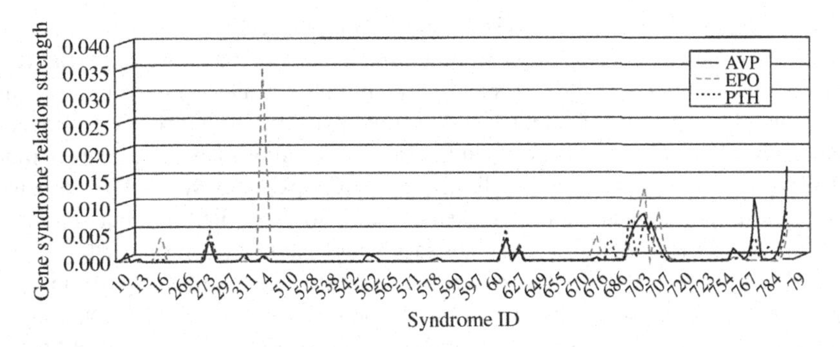

Figure 2.9 Gene relation strength distributions on syndromes. The figure displays the comparative distributions of three genes namely AVP, EPO, and PRL.

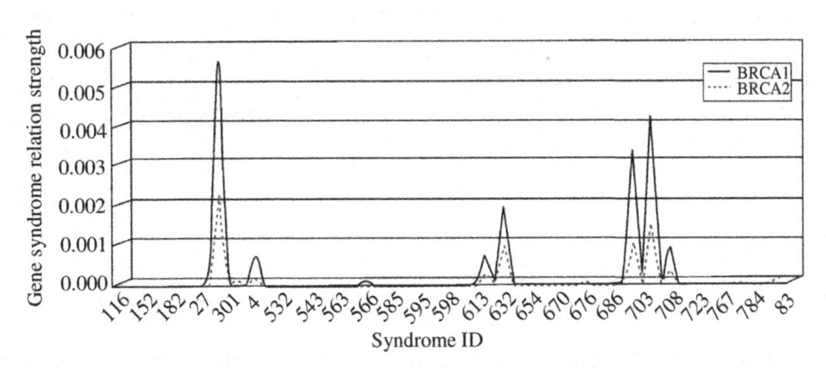

Figure 2.10 Gene relation strength distributions on syndromes. The figure displays the comparative distributions of the two genes namely BRCA1 and BRCA2.

2.7 Conclusions

Most biomedical literature mining studies are focused on the modern biomedical knowledge and the modern biomedical fields. In this chapter, we propose an integrative text-mining approach to generating large-scale biomedical knowledge using TCM holistic knowledge. We provide an approach to facilitating a marriage between the two independently developing and complementary sciences (i.e., modern biomedical science and TCM science). Hence, while introducing the syndrome knowledge to gene network research, we can generate lots of novel hypotheses for systems biology experimental efforts. We have developed the MeDisco/3S system to integrate and mine the data sources (e.g., MEDLINE and TCM bibliographic literature) from the two fields for functional gene knowledge. We have discussed the syndrome-oriented gene networks and gene functional analysis, which is worthy of further investigation. We believe that much more valuable hypotheses will be generated by enhancing the synergy of TCM and modern life sciences.

Currently, we use bubble-bootstrapping, term co-occurrence, and a pointwise mutual information-based relation weight computing method to generate the relation knowledge. As the ambiguities and polysemy of terminology used in literature, noise, and false-positive examples will surely exist in the currently discovered relation data. We will try to use a multi-strategy learning method to extract more reliable gene–disease or protein–disease relationships from MEDLINE. Because many new terminological names are used in the biomedical literature but are absent from the biomedical dictionary, a dictionary-based name extraction method is not applicable to extract the latest relationships. Therefore, we will study an unsupervised large-scale information extraction method based on bubble-bootstrapping to detect up-to-date relations from MEDLINE. Improving the relation extraction recall is also an important task for further research. We will build more complete multi-language (mainly in English and Chinese) disease terminology and integrate different information extraction methods.

As protein–protein interactions are central to most biological processes, the systematic identification of all protein interactions is considered a key to uncover the inner workings of a cell. The future work will also focus on extracting and discovering the protein–protein interactions by using integrative text-mining methods and TCM literature.

References

[1] Ideker T, Galitski T, Hood L. A new approach to decoding life: systems biology. Annu Rev Genomics Hum Genet 2001;2:343–72.

[2] Hatzivassiloglou V, Duboue PA, Rzhetsky A. Disambiguating proteins, genes and RNA in text: a machine learning approach. Bioinformatics 2001;17(Suppl. 1):S97–106.

[3] Fukuda K, Tsunoda T, Tamura A, et al. Toward information extraction: identifying protein names from biological papers. In: Altman RB, Dunker AK, Hunter L, et al, editors. Pacific symposium on biocomputing; 1998. p. 707–18.

[4] Stephens M, Palakal M, Mukhopadhyay S, et al. Detecting gene relations from MEDLINE abstracts. In: Altman RB, Dunker AK, Hunter L, et al, editors. Pacific symposium on biocomputing; 2001. p. 483–95.

[5] Marcotte EM, Xenarios L, Eisenberg D. Mining literature for protein–protein interactions. Bioinformatics 2001;17(4):359–63.

[6] Daraselia N, Yuryev A, Egorov S. Extracting human protein interactions from MEDLINE using a full-sentence parser. Bioinformatics 2004;20:604–11.

[7] Hao Y, Zhu X, Huang M, et al. Discovering patterns to extract protein–protein interactions from the literature. Bioinformatics 2005;21(15):3294–300.

[8] Blaschke C, Andrade MA, Ouzounis C, et al. Automatic extraction of biological information from scientific text: protein–protein interactions. In: Lengauer T, Schneider R, Bork P, et al., editors. Proceedings of the international conference on intelligent systems for molecular biology. Menlo Park, CA: AAAI Press; 1999. p. 60–7.

[9] Jenssen TK, Laegreid A, Komorowski J, et al. A literature network of human genes for high-throughput analysis of gene expression. Nat Genet 2001;28:21–28.

[10] Scherf M, Epple A, Werner T. The next generation of literature analysis: integration of genomic analysis into text mining. Brief Bioinform 2005;6(3):287–97.
[11] Hoeglund A, Blum T, Brady S, et al. Significantly improved prediction of subcellular localization by integrating text and protein sequence data. In: Altman RB, Murray T, Klein TE, et al., editors. Pacific symposium on biocomputing; 2006. p. 16–27.
[12] Eskin E, Agichtein E. Combining text mining and sequence analysis to discover protein functional regions. In: Altman RB, Dunker AK, Hunter L, et al., editors. Pacific symposium on biocomputing; 2004. p. 288–99.
[13] Glenisson P, Mathys J, De Moor B. Meta-clustering of gene expression data and literature-extracted information. ACM SIGKDD Explor Spec Issue Microarray Data Min 2003;5(2):101–12.
[14] Feng Y, Wu Z, Zhou X, et al. Knowledge discovery in traditional Chinese medicine: state of the art and perspectives. Artif Intell Med 2006;38(3):219–36.
[15] Cornish-Bowden A, Cárdenas ML. Systems biology may work when we learn to understand the parts in terms of the whole. Biochem Soc Trans 2005;33(3):516–19.
[16] Beijing University of Traditional Chinese Medicine. Basic theories of traditional Chinese medicine. Beijing: Academy Press; 2002.
[17] Yin HH, Zhang BN. The basic theory of traditional Chinese medicine (in Chinese). Shanghai: Shanghai Science and Technology Publishers; 1984.
[18] Pan M, Li L, Li K. Study on syndrome characteristics of Chinese medicine and relative factors in patients with DM. J Beijing Univ Trad Chin Med (Clin Med) 2004;13(4):6–10.
[19] The Department of Science Education, State Administration of Traditional Chinese Medicine of the People's Republic of China. The summary of workshop on traditional Chinese medicine and genomics (in Chinese). World Sci Technol/Modern Tradit Chin Med 1999;1:67–68.
[20] Shen ZY. The continuation of kidney study. Shanghai: Shanghai Science and Technology Publishers; 1990.
[21] Zhong LY, Shen ZY, Cai DF. Effect of three kinds (tonifying kidney, invigorating spleen, promoting blood circulation) recipes on the hypothalamus–pituitary–adrenal–thymus (HPAT) axis and CRF gene expression. Chin J Integr Med 1997;17(1):39–41.
[22] Li S, Zhang ZQ, Wu LJ, et al. Understanding ZHENG in traditional Chinese medicine in the context of neuro-endocrine-immune network. IET Syst Biol 2007;1:51–60.
[23] Mou H. Advancement of the treatment method proposed by Zhang Zhong-jing (in Chinese). Med Philos 2006;27(1):51–60.
[24] Chen K, Song J. Clinical study by way of combining diseases with differentiation of their syndromes, an important mode in study on combination of Chinese traditional and Western medicine. World Sci Technol/Modern Tradit Chin Med Mater Med 2006;8(2):1–5.
[25] Swanson DR. Two medical literatures that are logically but not bibliographically connected. J Am Soc Inf Retrieval 1987;38(4):228–33.
[26] Swanson DR. Complementary structures in disjoint science literatures. In: Bookstein A, Chiaramella Y, Salton G, et al., editors. Proceedings of the 14th annual international ACM SIGIR conference on research and development in information retrieval. New York, NY: ACM Press; 1991. p. 280–9.
[27] Swanson DR, Smalheiser NR. An interactive system for finding complementary literatures: a stimulus to scientific discovery. Artif Intell Med 1997;91(2):183–203.
[28] Gordon MD, Lindsay RK. Toward discovery support systems: a replication, re-examination, and extension of Swanson's work on literature-based discovery of a connection between Raynaud's and fish oil. J Am Soc Inf Sci 1996;47(2):116–28.

[29] Lindsay RK, Gordon MD. Literature-based discovery by lexical statistics. J Am Soc Inf Sci 1999;50(7):574–87.

[30] Weeber M, Klein H, Aronson AR, et al. Text-based discovery in biomedicine: the architecture of the DAD-system. In: Overhage JM, editor. Proceedings of the AMIA annual symposium. Philadelphia, PA: Hanley & Belfus; 2000. p. 903–7.

[31] Tanabe L, Xie N, Thom LH, et al. GENETAG: a tagged corpus for gene/protein named entity recognition. BMC Bioinf 2005;6(Suppl. 1):S3.

[32] Zhou G, Shen D, Zhang J, et al. Recognition of protein/gene names from text using an ensemble of classifiers. BMC Bioinf 2005;6(Suppl. 1):S7.

[33] Finkel J, Dingare S, Manning CD, et al. Exploring the boundaries: gene and protein identification in biomedical text. BMC Bioinf 2005;6(Suppl. 1):S5.

[34] Hakenberg J, Bickel S, Plake C, et al. Systematic feature evaluation for gene name recognition. BMC Bioinf 2005;6(Suppl. 1):S9.

[35] McDonald R, Pereira F. Identifying gene and proteinmentions in text using conditional random fields. BMC Bioinf 2005;6(Suppl. 1):S6.

[36] Xu H, Fan JW, Hripcsak G, et al. Gene symbol disambiguation using knowledge-based profiles. Bioinformatics 2007;23(8):1015–22.

[37] Hu X, Wu DD. Data mining and predictive modeling of biomolecular network from biomedical literature databases. IEEE/ACM Trans Comput Biol Bioinform 2007;4(2):251–63.

[38] Liu Y, Navathe SB, Civera J, et al. Text mining biomedical literature for discovering gene-to-gene relationships: a comparative study of algorithms. IEEE/ACM Trans Comput Biol Bioinform 2005;2(1):62–76.

[39] Bunescu R, Mooney R, Ramani A, et al. Integrating co-occurrence statistics with information extraction for robust retrieval of protein interactions from Medline. In: Proceedings of the HLT-NAACL workshop on linking natural language processing and biology: towards deeper biological literature analysis (BioNLP-2006); 2005. p. 49–56.

[40] Stapley BJ, Kelley LA, Sternberg MJ. Predicting the sub-cellular location of proteins from text using support vector machines. In: Altman RB, Dunker AK, Hunter L, et al, editors. Pacific symposium on biocomputing; 2002. p. 374–85.

[41] Rindflesch TC, Rayan JV, Hunter L. Extracting molecular binding relationships from biomedical text. Proceedings of the sixth conference on applied natural language processing. San Francisco, CA: Morgan Kaufmann Publishers; 2000. p. 188–195

[42] Rindflesch TC, Tanabe L, Weinstein JN, et al. EDGAR: extraction of drugs, genes and relations from the biomedical literature. In: Altman RB, Dunker AK, Hunter L, et al, editors. Pacific symposium on biocomputing; 2000. p. 517–28.

[43] van Driel MA, Cuelenaere K, Kemmeren PP, et al. GeneSeeker: extraction and integration of human disease-related information from web-based genetic databases. Nucleic Acids Res 2005;33:W758–761.

[44] van Driel MA, Bruggeman J, Vriend G, et al. A text-mining analysis of the human phenome. Eur J Hum Genet 2006;14(5):535–42.

[45] Chun HW, Tsuruoka Y, Kim JD, et al. Extraction of gene–disease relations from Medline using domain dictionaries and machine learning. In: Altman RB, Murray T, Klein TE, et al. editors. Pacific symposium on biocomputing; 2006. p. 4–15.

[46] Masseroli M, Galati O, Pinciroli F. GFINDer: genetic disease and phenotype location statistical analysis and mining of dynamically annotated gene lists. Nucleic Acids Res 2005;33(suppl 2):W717–W723. doi: 10.1093/nar/gki454.

[47] Chen Y, Shen C, Sivachenko AY. Mining Alzheimer disease relevant proteins from integrated protein interactome data. In: Altman RB, Murray T, Klein TE et al, editors. Pacific symposium on biocomputing; 2006. p. 367–78.

[48] Perez-Iratxeta C, Bork P, Andrade MA. Association of genes to genetically inherited diseases using data mining. Nat Genet 2002;31(3):316–9.
[49] Perez-Iratxeta C, Bork P, Wjst M, et al. G2D: a tool for mining genes associated with disease. BMC Genet 2005;6:45.
[50] Freudenberg J, Propping P. A similarity-based method for genome-wide prediction of disease-relevant human genes. Bioinformatics 2002;18(Suppl. 2):S110–115.
[51] Antal P, Fannes G, Timmerman D, et al. Using literature and data to learn Bayesian networks as clinical models of ovarian tumors. Artif Intell Med 2004;30(3):257–81.
[52] Bunescu R, Ge R, Kate RJ, et al. Comparative experiments on learning information extractors for proteins and their interactions. Artif Intell Med 2005;33(2):139–55.
[53] de Bruijn B, Martin J. Getting to the (c)ore of knowledge: mining biomedical literature. Int J Med Inform 2002;67(1-3):7–18.
[54] Hirschman L, Park JC, Tsujii J, et al. Accomplishments and challenges in literature data mining for biology. Bioinformatics 2002;18(12):1553–61.
[55] Hersh W. Evaluation of biomedical text-mining systems: lessons learned from information retrieval. Brief Bioinform 2005;6(4):344–56.
[56] Yandell MD, Majoros WH. Genomics and natural language processing. Nat Rev Genet 2002;3(8):601–10.
[57] Wilkinson D, Huberman BA. A method for finding communities of related genes. Proc Natl Acad Sci 2004;101(Suppl. 1):5241–8.
[58] Adamic LA, Wilkinson D, Huberman BA, et al. A literature based method for identifying gene–disease connections. Proceedings of the IEEE computer society bioinformatics conferences, 1. Washington, DC: IEEE Computer Society; 2002. p. 109–17.
[59] Yarowsky D. Unsupervised word sense disambiguation rivaling supervised methods. Proceedings of the 33rd annual meeting of the association for computational linguistics. Morristown, NJ: Association for Computational Linguistics; 1995. p. 189–96.
[60] Blum A, Mitchell T. Combining labeled and unlabeled data with co-training. Proceedings of the 11th annual conference on computational learning theory. New York, NY: ACM Press; 1998. p. 92–100.
[61] Jones R, McCallum A, Nigam K, et al. Bootstrapping for text learning tasks. In: Feldman R, editor. Proceedings of the 16th international joint conference on artificial intelligence workshop on text mining: foundations, techniques and applications. San Francisco, CA: Morgan Kaufmann; 1999. p. 52–63.
[62] Riloff E, Jones R. Learning dictionaries for information extraction by multi-level bootstrapping. Proceedings of the 16th national conference on artificial intelligence and the 11th innovative applications of artificial intelligence conference. Menlo Park, CA: AAAI Press; 1999. p. 474–9.
[63] Brin S. Extracting patterns and relations from the World Wide Web. In: Paolo A, Alberto M, Giansalvatore M, editors. Proceedings of the international workshop on the World Wide Web and databases LNCS 1590. London: Springer-Verlag; 1998. p. 172–183.
[64] Craven M, DiPasquo D, Freitag D, et al. Learning to extract symbolic knowledge from World Wide Web. Proceedings of the 15th national conference on artificial intelligence. Menlo Park, CA: AAAI Press; 1998. p. 509–16.
[65] Chien L, Pu H. Important issues on Chinese information retrieval. Comput Linguist Chin Lang Process 1998;1(1):205–21.
[66] Wu Z, Zhou X, Liu B, et al. Text mining for finding functional community of related genes using TCM knowledge. In: Boulicaut JF, Esposito F, Giannotti F, et al., editors.

Proceedings of the 8th European conference on principles and practice of knowledge discovery in databases LNAI 3202. Berlin: Springer-Verlag; 2004. p. 459–70.

[67] Zhou X. Issues in TCM text mining (in Chinese). PhD thesis, Zhejiang University; 2004.

[68] Lee I, Date SV, Adai AT, et al. A probabilistic functional network of yeast genes. Science 2004;306(5701):1555–8.

3 MapReduce-Based Network Motif Detection for Traditional Chinese Medicine[1]

3.1 Introduction

Network motifs are specific patterns of local interconnections with potential functional properties and can be seen as the basic building blocks of complex networks [1]. Pattern finding in a complex network is the first and most important step to analyze motifs. Some pattern-finding methods are already used to analyze the network motifs in the real world such as the biochemistry network, ecology network, neurobiology network, and engineering network [1]. And these applications provide many valuable research results. However, in the pattern-finding area, there are many NP-complete problems, such as determining graph isomorphisms and maximum independent sets [2]. For this reason, the pattern-finding algorithms always have high time–space complexity. Moreover, when the size of the pattern is big (usually bigger than 4), the number of the intermediate becomes very large (above millions of items), which makes pattern finding time consuming and memory exhausting.

Google's MapReduce framework is known as the framework of Cloud Computing. MapReduce built on top of the distributed Google File System provides a parallelization framework that has garnered considerable acclaim for its ease of use, scalability, and fault tolerance [3]. Therefore, we try to use the Google's MapReduce framework to speed up pattern finding and avoid running out of memory in a PC-cluster environment. We design a MapReduce-based pattern-finding (MRPF) algorithm that provides good efficiency and scalability. We also apply it to a prescription network and successfully find some commonly used prescription structures that propose the possibility to discover the law of prescription compatibility.

In our MRPF algorithm, we reorganize the traditional pattern-finding process into four steps: distributed storage, neighbor vertices finding and pattern initialization, pattern extension, and frequency computing. Each step is implemented by a

[1] With kind permission from Liu, Y, Jiang, X, Chen, H, Ma, J, Zhang, X. MapReduce-based pattern finding algorithm applied in motif detection for prescription compatibility network. Springer Science+Business Media: APPT 2009; 2009: 341–55.

Modern Computational Approaches To Traditional Chinese Medicine. DOI: http://dx.doi.org/10.1016/B978-0-12-398510-1.00003-0

MapReduce pass. In each MapReduce pass, the task is divided into a number of subtasks of the same size and each subtask is distributed to a node of the cluster. MRPF uses an extended mode to find the target-size pattern, i.e., trying to add one more vertex to the matches of i-size patterns to create patterns of size $i + 1$. The extension does not stop until patterns reach the target size.

To test the computational efficiency of MRPF, we apply it to the prescription compatibility structure detection. The knowledge discovery of prescription compatibility is an important part of traditional Chinese medicine (TCM) research. Prescription compatibility investigates the composite structure of herbal medicines. One prescription contains five or six herbal medicines. However, prescriptions are commonly given based on experiences without theoretical instruction on prescription structures. So we construct the prescription compatibility network and use our algorithm to detect the prescription compatibility structure.

The rest of the chapter is organized as follows. Section 3.2 introduces some related works on pattern-finding methods, applications of MapReduce, and some data mining methods used in TCM. Section 3.3 describes our MRPF algorithm in detail. Section 3.4 gives a case study on prescription compatibility using our algorithm. Section 3.5 provides some concluding remarks and discussion for future work.

3.2 Related Work

The main step in motif detection is pattern finding in the complex network. There are two distinct problem formulations for pattern finding in graph datasets. One is the graph-transaction setting that uses a set of relatively small graphs as input data, the other is the single-graph setting using a single large graph instead [4]. Pattern finding in graph-transaction attracts more attention, so that a number of efficient algorithms [5–10] have been developed. However, few investigations have been made in pattern finding from the single-graph setting. Moreover, some algorithms, such as GBI [11] and SUBDUE [12], will lose a large number of patterns, and at the same time not scale well for large datasets due to computational complexity. In recent years, with the application of pattern finding increasingly used in many fields, researchers start to pay more attention to designing algorithms for single-graph setting. In 2005, Kuramochi and Karypis [4] developed an algorithm to find patterns in a large sparse graph. Schreiber and Schwöbbermeyer [13] designed an FPF multiple-thread algorithm to improve the performance of Michihiro's algorithm. Chen et al. [14] designed a NeMoFinder algorithm that can mine mesoscale network motifs in large protein–protein interaction networks. In 2007, Chen et al. [15] invented a gApprox algorithm that does consider approximate matching in its search space. However, all these algorithms mentioned above failed to consider the limitation of the main memory of one computer. So for further improving the performance of pattern finding and breaking through single computer resource constraints, we design a parallel pattern-finding algorithm based on MapReduce framework. It is a complete algorithm without losing any target-size pattern in the network. MapReduce framework, as a parallel model, is often used in data mining, such as machine learning [16] and SVM [17]. These experiments demonstrate that the MapReduce framework is effective for problems with high complexity and a large dataset. It is also proved that MapReduce can be adapted to manipulating graphs. Implementation of pattern finding in the context of MapReduce framework is able to address the issues of insufficient memory, computational complexity, and fault tolerance. Many data mining methods have been used in the modernization of TCM. A text-mining method is used for finding a functional community of TCM knowledge [18]. Mining compatibility rules are used for TCM databases [19]. A clustering method is applied to analyze Chinese text categorization [20]. Prescription compatibility is investigated in Refs. [21–23], but very little work has been done on motif detection in the prescription network, which is very important for law discovery of prescription compatibility. We analyze the prescription compatibility using the complex network and find some commonly used prescription structures.

3.3 MapReduce-Based Pattern Finding

MRPF framework aims to implement frequent pattern finding on complex graphs based on Hadoop. Although it also works well on undirected graphs, here we still focus on introducing its application to directed graphs. It is more interesting and representative to apply this framework on directed graphs. For clearly depicting MRPF, we show the serial pattern-finding algorithm in Algorithm 3.1.

3.3.1 MRPF Framework

Here we define the size of a pattern by its vertices number. We use the generation of a canonical label described in Ref. [24] to check graphs for isomorphism. After loading a dataset of a network, MRPF uses one MapReduce pass to parse the dataset and form three information tables. Another MapReduce pass is used to extend matches that are subgraphs of the network from size i to $i + 1$. The frequency of new patterns will be calculated after all matches of patterns of size $i + 1$ have been obtained. Figure 3.1 depicts the outline of MRPF.

Algorithm 3.1 Normal Pattern Finding

```
Data: Dataset of Graph G, target pattern size s, minimum support (f_min)
Result: Set P of pattern of target size begin
P←{all pattern of size 2};
size←2;/* initial size */
MATCHp2←all matches of p2;
TPS←φ/* TPS: target pattern size */
while size<target size do
  foreach pattern p ∈ P do
    foreach match m ∈ MATCHp do
```

```
        foreach incident vertex v of m do
          m'←m ∈{v};
          p'←pattern of {m'};
          TPS←TPS ∪ {p};
          MATCHp'←MATCHp'∪ {m'};
        end
      end
    end
  p←φ;
  foreach p ∈ TPS do
    frequency←sizeof (MATCHp);
    if frequency > f_min then P←P∪ {p};
  end
  size++;
end.
```

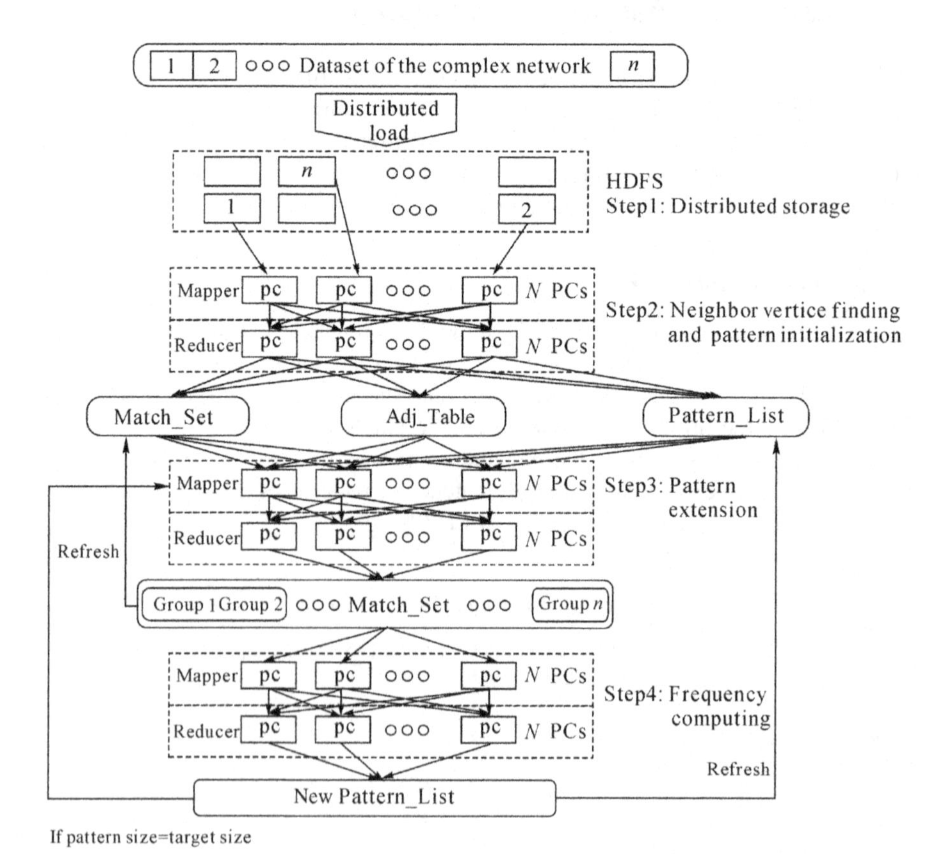

Figure 3.1 The MRPF framework. HDFS, Hadoop Distributed File System.

Step 1: *Distributed storage.* MRPF is based on Hadoop, a Google File System implementation, hosted as a project of the Apache Software Foundation [16]. In Step 1, the target network is stored as textual files in a specific format. Using Hadoop, the file can be divided easily into a set of blocks with the same size and distributed on nodes of the cluster to keep load balance in the cluster. Hadoop can process the blocks concurrently on nodes where the data is located.

Step 2: *Neighbor vertices finding and pattern initialization.* In this step, we use a MapReduce pass to do two tasks: one is to find an adjacent neighbor of each vertex to form an adjacent vertices table (Adj_Table) and the other is to find patterns of size two (one edge and two vertices) and their matches. Each mapper inputs one block of the dataset. The results are respectively stored in Adj_Table, Match_Set, and Pattern_List. Note that Adj_Table is distributed to every node in the cluster and it will be used in pattern extension (Step 3). It is used to detect the patterns on the borders of blocks and to guarantee our algorithm to be complete (against losing patterns). Match_Set and Pattern_List are updated by Steps 3 and 4, respectively. We will introduce the details in Section 3.3.2.

Step 3: *Pattern extension.* It is the key step of the MRPF. This step also takes one MapReduce pass. The map stage working with reduce stage extends patterns of size i to $i+1$. The details will be explained in Section 3.3.3.

Mapper—extends the matches of size i to $i+1$, calculates their patterns, and produces a group key with the patterns and matches. Each mapper outputs one or more key-value pairs, and the pairs with the same key will automatically be grouped into the same reducer.

Reducer—removes the duplicated matches, because different matches may get the same subgraph of size $i+1$ when the matches i are extended. During the grouping process mentioned above, we compare the canonical label of each match and keep just one of the same matches. The outputs of reducers are grouped into different files according to the pattern label.

Step 4: *Frequency computing.* After pruning the identical subgraphs, a MapReduce pass is used to count the support value of all patterns that appear in the big simple graph. We prune the patterns lower than the minimum required frequency. Then we store new patterns in Pattern_List. Then we go back to Step 3 to process iteratively until the target pattern size is reached. The details are given in Section 3.3.4.

3.3.2 Neighbor Vertices Finding and Pattern Initialization

Just like a classical application of MapReduce, each mapper of the first MapReduce pass is fed with one block of dataset. The input key-value pairs would be like <key, value=Edge(V_i,V_j)> (V_i and V_j are adjacent vertices to each other), where edges belong to dataset. Mappers produce two kinds of keys: the vertex key according to vertex label and the pattern key according to the pattern canonical label. Mappers travel through all edges of the graph, each mapper outputs three key-value pairs <$key_1=V_i$, $value_1=V_j$>, <$key_2=V_j$, $value_2=V_i$> and <$key_3=pattern_2$, $value_3=$Edge(V_i,V_j)>.

After all mapper instances have finished, the MapReduce infrastructure automatically collects and groups the key-value pairs according to the keys. The values with the same key are put into the same group, called *G*(key) and reducers receive the key-value pairs <key, *G*(key)> where *G*(key) is adjacent vertices of a vertex or the match of a pattern whose size is two. Reducers compose the *G*(key) into an adjacent vertices list or match list and output <key, list> into Adj_Table or Match_Set according to the class of each key, where the list is a vertices list or match list. Algorithm 3.2 presents the pseudo code of this step. Through this process, it registers each vertex's adjacent vertices. Meanwhile, it finds the smallest patterns (of size 2) and their matches.

Algorithm 3.2 Neighbor Vertices Finding and Pattern Initialization

```
Procedure: Mapper(key, value=Edge(Vi:Vj))
/* p is the canonical label of Edge(Vi:Vj) */
p getPattern(Edge(Vi:Vj))
EmitIntermediate (<key=Vi, value=Vj>)
EmitIntermediate(<key=Vj, value=Vi>)
EmitIntermediate (<key=p, value=Edge(Vi:Vj)>)
Procedure: Reduce(key, value=G(key))
/* Adj_List:adjacent vertices list;
  Match_List:matches of the same pattren */
Adj_List
Match_List
if key is vertex label then
  foreach item vi in G(key) do
     Adj_List Adj_List {vi}
  end
  Emit('<key, Adj_List> )
else
  foreach item match in G(key) do
     Match_List Match_List {matchi}
  end
  Emit( <key, Match_List> )
end
```

3.3.3 Pattern Extension

This step is the key part of the MRPF algorithm. This step, together with Step 4, will be repeated until the target-size pattern is obtained. In the pseudo code of Algorithm 3.3, we will see the procedure of how we use the MapReduce framework clearly.

First, we load the adjacent vertices table (Adj_Table) which can be stored in the memory to find incident vertices of matches. Then we load the Pattern_List which keeps all the patterns of size *i*. The initial state of the Pattern_List is defined as "Starting." The input of mapper is from Match_Set. As shown in Figure 3.2, if a match (highlighted in bold black line) is found in the graph, we call the vertices (in gray) adjacent to the matched vertices (in black) the incident vertex of a match. And the edge between an incident vertex and the match is called the detected edge.

In mappers, the input pair would be like <key, value=Pattern$_i$ & match$_i$> from the Match_Set, where Pattern means the match has i vertices and match$_i$ means the match has i vertices. If the Pattern$_i$ is contained in Pattern_List or Pattern_list is in initial state, it extends match$_i$ to match$_{i+1}$ through adding each incident vertex into match$_i$, as shown in Figure 3.3. Then it adds various combinations of detected edges into match$_i$ and forms new matches.

We compute Pattern$_{i+1}$ of match$_{i+1}$ and the canonical label of the Pattern$_{i+1}$ as part of output value. Each mapper outputs one or more pairs like <key′=match$_{i+1}$, value′=pattern$_{i+1}$>.

Each reducer receives a <key′=match, S (key′)>, where S (key′) has only one element—the pattern of the match$_{i+1}$ that is the key of key-value pair. In this way, we can easily wipe off the identical matches. Each reducer outputs one <key″=Pattern$_{i+1}$, key′>, the MapReduce infrastructure sorts and groups the key-value pairs according to the key″ value, then produces the successive block based on grouping. Blocks are stored in N different computers, which is convenient to deal with the data in the next MapReduce process.

Algorithm 3.3 Pattern Extension

```
Procedure: Mapper(key, value=Patterni & matchi)
  Load Adj_Table
  Load Pattern_List
if Patterni in Pattern_List
  or Pattern_list is in the Starting state then
  foreach incident vertex of matchi do
    foreach combination Ci of detected edges
      between the incident vertex and matchi do
      matchi+1 matchi {Ci}
      Patterni+1 corresponding pattern of matchi+1
        EmitIntermediate(<key=matchi+1, Patterni+1>)
    end
  end
Procedure: Reduce(key= matchi+1, S (key))
  Patterni+1 one of S (key)
  Emit(<key=Patterni+1, matchi+1>)
```

3.3.4 Frequency Computing

Frequency computing is a simple counting process, a classical application of MapReduce. Its input is the output of Step 3. Algorithm 3.4 presents the pseudo

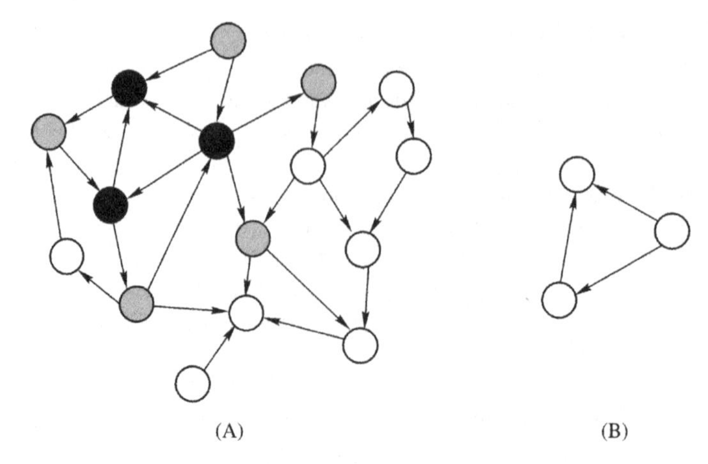

Figure 3.2 (A) GT is a graph. The subgraph (highlighted with bold lines) in GT is a match of the pattern in (B). Gray vertices in GT are incident vertices of the match and gray edges are detected edges. (B) A pattern of size 3.

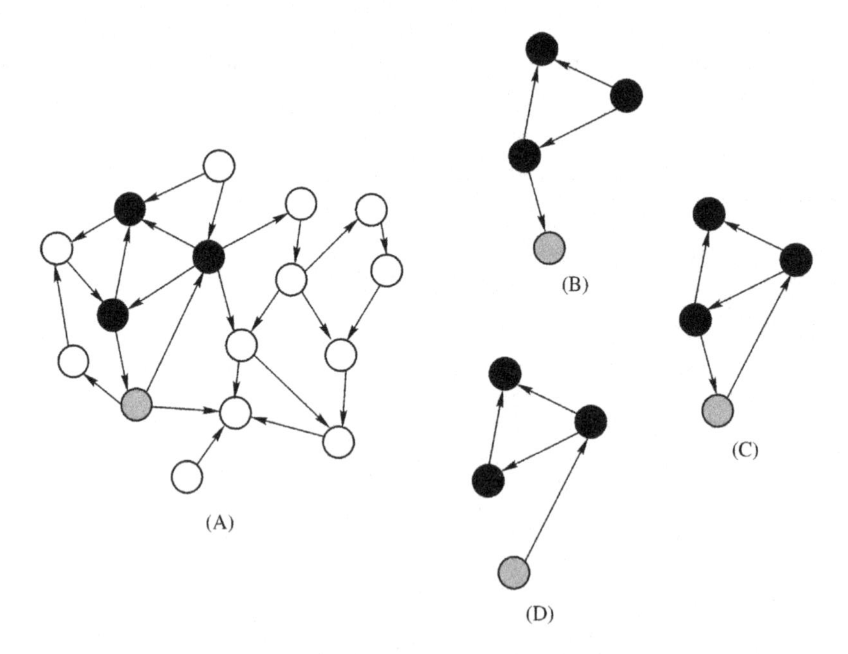

Figure 3.3 (A) A graph with a randomly selected subgraph (highlighted with bold lines). The subgraph is a match (M3) of some pattern. (B)–(D) are the extension matches from M3.

code of the two steps: grouping and parallel counting. The mapper's input is <key, value = T>, where T is composed of a pattern canonical label and matches. It picks up the pattern canonical (P) from T. The mapper outputs a key-value pair <key′ = P, value′ = 1>.

After completing all of the Mapper instances, for each key transmitted by Mapper, a value set (*S* (key′)) is automatically formed. And each reduce is fed with <key′, *S* (key′)>. The reducer outputs <key″ = key′, value″ = sum(*S* (key′))>.

Then the pattern frequency is calculated based on the occurrence quantity of each pattern. In this chapter, to show the full potential of the prescription network, we use the frequency concept which counts every match of the pattern. It gives a complete overview of all possible occurrences of a pattern, even if elements of the target graph are used several times. Thus it does not satisfy the downward closure property [4]. And we do not prune the infrequent patterns that are lower than the target pattern size. Note that the occurrence quantity of some patterns is too small to affect the pattern-finding results. We call these patterns dust patterns. A minimum required frequency variable (f_min) is defined to prune the dust patterns. The value of f_min is given by experts according to their experience.

Algorithm 3.4 Frequency Computing

```
Procedure: Mapper(key, value=Patterni & matchi)
/* PL: pattern label*/
PL the canonical label of Patterni
  EmitIntermediate(<key=PL, '1'>)
Procedure: Reduce(key, S (key))
  Sum 0
foreach item '1' in S(key) do
  Sum Sum+1
End
/* total is the quantity of all patterns, f_min is a minimum required frequency
variable */
if Sum total * f_min
  Emit(<key = Patterni+1, sum>)
end
```

3.4 Application to Prescription Compatibility Structure Detection

Because we have described the MRPF algorithm in detail, in this section, some case studies on prescription compatibility using the MRPF algorithm will be shown.

3.4.1 Motifs Detection Results

A key subject of prescription research is theoretical study of prescription compatibility regularity. The structure of Monarch, Minister, Assistant, and Guide is the compatibility principle for prescriptions and the base for the overall efficacy of

Table 3.1 All Types of Possible Compatibility Relations

Herbal	Compatibility Relation	Herbal	Compatibility Relation
Monarch, Minister	Monarch→Minister	Monarch, Assistant	Monarch→Assistant
Monarch, Guide	Monarch→Guide	Minister, Assistant	Minister→Assistant
Minister, Guide	Minister→Guide	Assistant, Guide	Assistant→Guide

prescriptions. However, people as yet know nothing about the commonly used compatibility structure. In other words, people still have no acquaintance with the most appropriate ratio of these four kinds (e.g., Monarchs, Ministers, Assistants, and Guides) of Chinese herbal medicines, respectively, participating in the compatibility of a prescription, which is very important in exerting the overall efficacy.

In the prescription compatibility network, a node represents Chinese herbal medicine, while an edge describes the compatibility relation that might exist between the two herb nodes, and edge direction indicates the relative position between the two connected herb nodes from the higher one to the lower one. According to the multi-types of the relative positions between any two herb nodes, the compatibility relations of them vary greatly. Table 3.1 shows in detail all types of possible compatibility relations in terms of criteria for the classification of herbs.

We select 201 prescriptions that are explicit in the compatibility structure from Ref. [25] to construct the prescription compatibility network. The network contains about 300 vertices and 2000 edges.

We apply our algorithm in the prescription network, and after comparing it with random networks, we find a number of motifs of the prescription network (Figure 3.4) and their occurrence quantity shown in Table 3.2.

These motifs are the basic structure of the prescription compatibility. For example, Motif 1 consists of one Monarch, one Minister, one Assistant, and one Guide; Motif 3 contains one Monarch, one Minister, one Assistant, and one Guide; Motif 5 contains one Monarch, two Ministers, and two Assistants. They are of great value to further discover the law of prescription compatibility.

3.4.2 Performance Analysis

Our algorithm automatically divides the job and distributes them to each node. Thus we can dynamically add the quantity of nodes, which will enhance the performance of the algorithm. We run the program on a blade-cluster with 48 nodes. Each node is equipped with Intel (R) Xeon (TM) CPU 2.80 GHz and 1 GB memory. In the experiment, we run the algorithm to do the same task on a cluster of varying nodes. The experiment results of finding size 4 and size 5 motifs are shown in Figure 3.5.

From the figure, it is clear that execution time decreases quickly while the cluster nodes increase. It implies that our algorithm scales well with the computing nodes. However, the performance acceleration decreases when the cluster exceeds

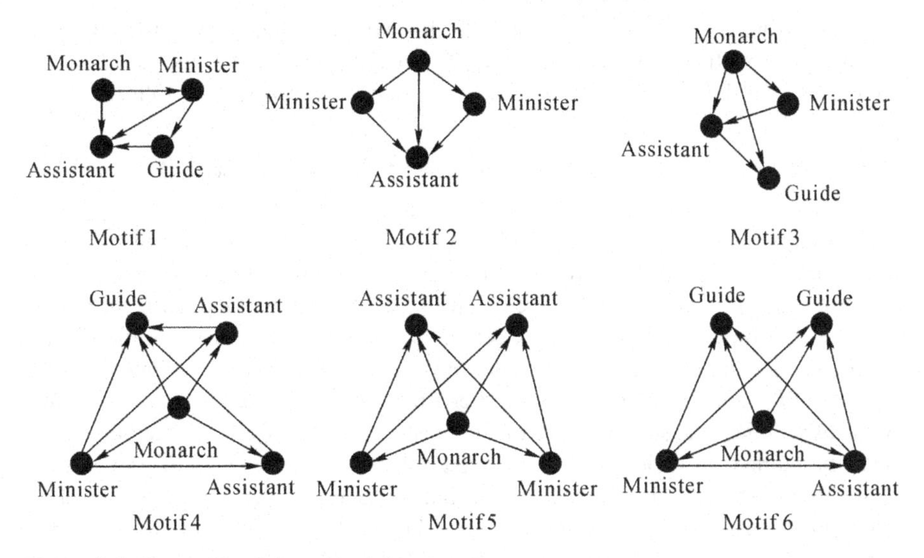

Figure 3.4 Six motifs of sizes 4 and 5 separately.

Table 3.2 Frequency for Each Motif in Figure 3.4

Motif	Frequency (%)
1	0.4093750
2	0.3534964
3	0.3728833
4	0.0038188
5	0.0026400
6	0.0018862

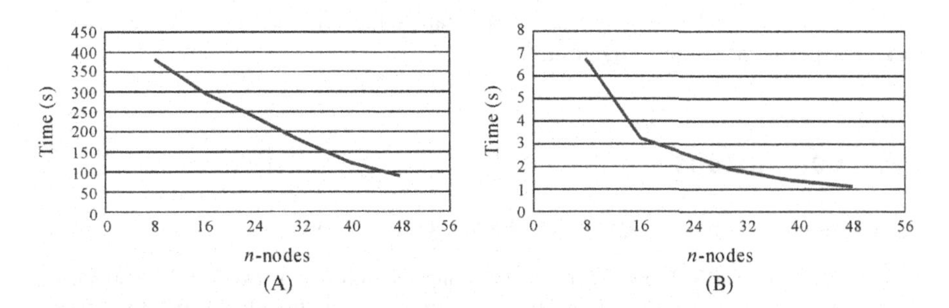

Figure 3.5 Algorithm performance on the cluster: (A) motif size = 4 and (B) motif size = 5.

a number of nodes for a fixed-size task. In this experimentation, we just prove the scalability of MRPF. Here we theoretically analyze reasons for the acceleration decreasing. We define the formula of the execution time of MRPF as follows:

$$T_{\text{total}}(N) = \frac{c_{\text{fix}}}{N} + T_{\text{overhead}} = \frac{c_{\text{fix}}}{N} + T_{\text{map}}(N) + T_{\text{reduce}}(N) + T_{M/S}(N) \quad (3.1)$$

where N is the number of data nodes, C_{fix} is a constant that represents the computation complexity of the fixed-size task and it is distributed to each node $T_{\text{map}}(N)$, $T_{\text{reduce}}(N)$, and $T_{M/S}(N)$ evenly; respectively denote Map Task initialization time, the time for each reducer receiving the intermediate, and the time for the communication between the master and slave nodes. The Map Task initialization time $T_{\text{map}}(N)$ includes assigning tasks, preparing data, and task issuing. According to our experience, the total number of map tasks is better to be set at about 3 or 4 times the number of nodes, which can make full use of the resource of the cluster. So we divide the task dynamically according to N. While the number of the data blocks increases with N, the number of the Map Task increases and the total time of map initialization and intermediate distribution increase too. It is obvious that $T_{\text{map}}(N)$ and $T_{\text{reduce}}(N)$ are increasing with N. And the cluster is organized in master/slaves mode, with only one master node responsible for data retrieval, task assignment, and task snooping on the slave nodes. So while the node number N increases, the communication overhead between master and slaves will also increase. So the value of $T_{M/S}(N)$ is increasing with N. The speedup of MRPF can be calculated using the following formula:

$$\begin{aligned} \text{Speedup} &= \frac{T_{\text{total}}(N)}{T_{\text{total}}(N+1)} \\ &= \frac{(C_{\text{fix}}/N) + T_{\text{map}}(N) + T_{\text{reduce}}(N) + T_{M/S}(N)}{(C_{\text{fix}}/(N+1)) + T_{\text{map}}(N+1) + T_{\text{reduce}}(N+1) + T_{M/S}(N+1)} \end{aligned} \quad (3.2)$$

From the above equation, it can be deduced that the speedup might degrade due to the increasing overhead, even if C_{fix} is allocated by the cluster nodes. It can be clearly observed that there is an inflection point in the exertion time curve in Figure 3.5B when the number of the data nodes equals 16. It may be caused by the topology of the cluster or architecture of MapReduce framework. We need to do further experiments to investigate it.

3.5 Conclusions

In summary, the contributions of this chapter are as follows:

1. We designed an MRPF algorithm for analyzing the complex network. We reorganized the pattern-finding process and implemented each step using the MapReduce framework, which makes MRPF parallelizable and extensible. The experiment evaluation on the

expending of nodes in Section 3.4.2 indicated that increasing the number of the nodes would enhance the performance of MRPF.

2. We applied the complex network analysis method to the prescription compatibility network and used MRPF to find the commonly used compatibility structure. And we found some prescription structures which reflect characteristics of the law of compatibility of medicines in prescriptions in some way.

More experiments need to be done to evaluate the algorithm performance considering the factors of data block size, node number, network bandwidth, etc. In fact, developing the MRPF algorithm is actually the first step to our target to develop a parallel data mining library based on MapReduce that can be applied in many fields. And we will also test these parallel algorithms in data mining in TCM.

References

[1] Milo R, Shen-Orr S, Itzkovitz S, et al. Network motifs: simple building block of complex networks. Science 2002;5594:824–7.

[2] Garey MR, Johnson DS. Computers and intractability: a guide to the theory of NP-completeness. New York, NY: W.H. Freeman and Company; 1979.

[3] Dean J, Ghemawat S. MapReduce: simplified data processing on large clusters. In: ACM OSDI; 2004. p. 137–50.

[4] Kuramochi M, Karypis G. Finding frequent patterns in a large sparse graph. In: Data mining and knowledge discovery. Heidelberg: Springer; 2005. 5810: p. 243–71.

[5] Yan X, Han J. gSpan: graph-based substructure pattern mining. In: 2002 IEEE international conference on data mining, 2002. ICDM 2002. Maebashi City: Proceedings IEEE Press; 2002. p. 721–4.

[6] Inokucbi A, Wasbio T, Motoda H. Complete mining of frequent patterns from graphs: mining graph data. Mach Learn 2003;50(3):321–54.

[7] Hong M, Zhou H, Wang W, et al. An efficient algorithm of frequent connected subgraph extraction. In: Whang KY, Jeon J, Shim K, et al., editors. PAKDD 2003. Heidelberg: Springer; 2003. LNCS 2637: p. 40–51.

[8] Yan X, Hart J. CloseGraph: mining closed frequent patterns. The 9th ACM SIGKDD international conference on knowledge discovery and data mining (KDD 2003). Washington, DC: ACM; 2003. p. 286–95.

[9] Huan J., Wang W., Prins J. Efficient mining of frequent subgraph in the presence of isomorphism. In: 2003 International conference on data mining (ICDM), Melbourne, IEEE, FL; 2003. p. 549–52.

[10] Gudes E, Shimony SE, Vanetik N. Discovering frequent graph patterns using disjoint paths. IEEE Trans Knowl Data Eng 2006;18(11):1441–56.

[11] Yoshida K, Motoda H, Indurkhya N. Graph-based induction as a unified learning framework. J Appl Intell 1994;4:297–328.

[12] Cook J, Holder L. Substructure discovery using minimum description length and background knowledge. J Artif Intell Res 1994;231–55.

[13] Schreiber F, Schwöbbermeyer H. Frequent concepts and pattern detection for the analysis of motifs in networks. In: Priami C, Merelli E, Gonzalez P, et al., editors. Transactions on computational systems biology III. Heidelberg: Springer; 2005. LNCS (LNBI) 3737: p. 89–104.

[14] Chen J, Hsu W, Lee ML, et al. Nemofinder: dissecting genome-wide protein–protein interactions with meso-scale network motifs. In: Proceedings of the 12th ACM SIGKDD international conference on Knowledge discovery and data mining. New York, NY; 2006. p. 106–115.
[15] Chen C, Yan X, Zhu F, et al. gApprox: mining frequent approximate patterns from a massive network. In: Perner P, editor. ICDM 2007. Heidelberg: Springer; 2007. LNCS (LNAI) 4597: p. 445–50.
[16] Chu C, Kim SK, Lin Y, et al. Map-reduce for machine learning on multicore. NIPS 2006;281–8.
[17] Chang E, Zhu K, Wang H, et al. PSVM: parallelizing support vector machines on distributed computers. NIPS 2007.
[18] Wu Z, Zhou X, Liu B, Chen J. Text mining for finding functional community of related genes using TCM knowledge. In: Boulicaut JF, Esposito F, Giannotti F, et al., editors. PKDD 2004. Heidelberg: Springer; 2004. LNCS (LNAI) 3202: p. 459–70.
[19] Tan Y, Yin GF, Li GB, et al. Mining compatibility rules from irregular Chinese traditional medicine database by apriori algorithm. J Southwest Jiaotong Univ 2007;15:288–92 [English edition].
[20] Zhou XZ, Wu ZH. Distributional character clustering for Chinese text categorization. In: Zhang C, Guesgen HW, Yeap WK, editors. PRICAI 2004. Heidelberg: Springer; 2004. LNCS (LNAI) 3157: p. 575–84.
[21] Xiao HB, Liang XM, Lu PC, et al. New method for analysis of Chinese herbal complex prescription and its application. Chin Sci Bull 1999;44:1164–72.
[22] Feng Y, Wu Z, Zhou X, et al. Knowledge discovery in traditional Chinese medicine: state of the art and perspectives. Artif Intell Med 2006;38(3):219–36.
[23] Chang YH, Lin HJ, Li WC. Clinical evaluation of the traditional Chinese prescription Chi–Ju–Di–Huang–Wan for dry eye. Phytother Res 2005;19(4):349–54.
[24] Kuramochi M, Karypis G. An efficient algorithm for discovering frequent subgraphs. Technical report 02-026, Department of Computer Science, University of Minnesota; 2002.
[25] Fujing D. Prescription: for the specialty of Chinese traditional medicine. Shanghai: Shanghai Publishing House of Science and Technology Press; 2006.

4 Data Quality for Knowledge Discovery in Traditional Chinese Medicine[1]

4.1 Introduction

The past few years have witnessed phenomenal growth of bioinformatics and medical informatics. Due to the wide availability of large-scale database systems and high-throughput instruments, huge amounts of data are generated and stored, not only in bioinformatics but also in medical informatics. Such tremendous growth of biomedical data, however, has exacerbated two existing problems. On the one hand, the fast-growing amount of data has far exceeded our human ability for comprehension without powerful tools. Consequently, more effective techniques and methods should be developed to transform such data into useful information and knowledge, which is known as the area of data mining and knowledge discovery. On the other hand, in spite of the biomedical data explosion, doctors and researchers still often face the lack of high-quality data that totally satisfies their needs. These two problems are actually closely related: data of high quality is indispensable for data-mining techniques to extract reliable and useful knowledge, and a knowledge discovery process in the real world typically includes steps to identify and treat data quality problems. In this chapter, we focus on the second problem, i.e., the data quality issue, and discuss how to handle data quality in a subarea of a medical field, namely traditional Chinese medicine (TCM).

The topic of data quality is covered in much literature. A substantial amount of work has been devoted to identifying the dimensions of data quality and their interrelationships. In 1987, Ballou and Pazer [1] identified some key dimensions of data quality: accuracy (the recorded value is in conformity with the actual value), timeliness (the recorded value is not out of date), completeness (all values for a certain variable are recorded), and consistency (the representation of the data value is the same in all cases). In 1995, Wang et al. [2] proposed a hierarchy of data quality dimensions and defined accessible, interpretable, useful, and believable as four basic dimensions in the hierarchy. A more comprehensive study was conducted by

[1] Feng Y, Wu, Z, Chen, H, Yum T, Mao, Y, Jiang, X. IEEE. Reprinted with permission from International Conference on BioMedical Engineering and Informatics; 2008. p. 268. doi:10.1109/BMEI.

Modern Computational Approaches To Traditional Chinese Medicine. DOI: http://dx.doi.org/10.1016/B978-0-12-398510-1.00004-2

Wang and Strong [3] in 1996. With a two-stage survey, they identified 179 attributes of data quality and then combined these attributes into a conceptual hierarchy framework of data quality with four basic dimensions (intrinsic quality, contextual quality, representation, and accessibility) and 15 subdimensions. A recent review about data quality in 2006 can be found in Ref. [4]. As indicated by Kulikowski [5], there is no universal set of data quality descriptors existing, and they should be chosen according to the application area specificity.

Among various application fields, health care and medicine is an area where data quality bears particular importance. As the basis of medical decision making, data is useful only when it is of good quality [6]. For diagnosis, prescribing, or other forms of health care activity, any decision making based on low-quality data might result in some unwanted, sometimes even disastrous consequences. Examples of the implications of poor-quality data in health care could be found in Ref. [7]. In the past few years, there has been a large amount of research discussing data quality in medical informatics [6–9]. A review of data quality literature in medical registries was given by Danielle and Nicolette [6] in 2002.

However, existing works on this aspect basically discuss the data quality issues in the context of orthodox medicine. To our knowledge, little data quality research is published in the background of complementary and alternative medicine (CAM), particularly in TCM. As a complete medical knowledge system, TCM has played an indispensable role in the health of Chinese people for several thousand years. More importantly, TCM has gained ever-increasing popularity in the last decades, not only in China, but also in other parts of the world. Countless TCM practices and theoretical research over thousands of years accumulated a great deal of knowledge that is included in ancient books, modern literature, and electronic health records. To make full use of these data and extract useful knowledge from these resources, our research group (CCNT Lab) and China Academy of Traditional Chinese Medicine (CATCM) have collaborated to build the TCM-online Database System [10] since 1998 under the support of the Chinese government. Currently, the TCM-online Database System is the largest TCM data collection in the world, integrating 17 branches in the whole country and more than 50 TCM-related databases. In recent years, as one of the designers and builders of the TCM-online Database System, a series of explorations has been carried out by CCNT Lab to conduct knowledge discovery in TCM data [10–17]. During these efforts, some experience about data quality in TCM has been accumulated and data quality is regarded as one of the key factors which could affect knowledge discovery reliability [16]. Considering the high domain specificity of the data quality issue and the poor quality of existing TCM data, it is necessary to discuss data quality problems in the TCM background. This chapter is an attempt to introduce the experience and methods accumulated in years of practice, which might be beneficial to TCM doctors and research with regard to the data quality issue.

The rest of this chapter is arranged as follows. Based on our real-world experience in past years, an introduction of three key data quality dimensions in TCM is given in Section 4.2. Section 4.3 follows with a presentation of methods and techniques to handle data quality problems in these dimensions. Finally, we conclude in Section 4.4.

4.2 Key Data Quality Dimensions in TCM

The scope of data quality dimensions to be considered varies according to applications. With years of experience of data collection, integration, and analysis in the TCM field, we regard the following data quality dimensions as worth more attention: representation granularity, representation consistency, and completeness. (Another significant data quality dimension in TCM is trustworthiness. However, due to space limitations, the discussion of trustworthiness in TCM will be given in another chapter.) It should be noted that the discussion of these dimensions here does not mean their superior importance over other dimensions. They are highlighted because most data quality problems in the TCM-online Database System belong to these dimensions. Consequently, we focus on these three dimensions in this section and introduce them one by one in the following sections.

4.2.1 Representation Granularity

This dimension means whether the data fields represent information in the right granularity. Actually, this type of data quality problem could possibly be avoided when the database model or data warehouse model is perfectly designed. However, this condition is hard to fulfill in the real world. The main problem of representation granularity we find in TCM data is multiple data elements in one data field. Ideally, there should be exactly one data field for one data element in each record. However, this condition is always violated in practice. Take the case illustrated in Ref. [18] as an example: a data field containing "Mr. Frank M." can be easily understood by a human operator. However, this value actually includes three fields: the title "Mr.," the first name "Frank," and the middle initial "M." Thus, if one wants to know the number of male customers, one might get an incorrect result if this data field is not treated. Such phenomena are very common in TCM databases. For instance, the ingredients, which refer to the Chinese herbal medicines constituting a Chinese medical formula, are stored as only one attribute in DCMF (Database of Chinese Medical Formula). Actually, this single data field contains not only the names of Chinese herbal medicines but also the corresponding dosages, weight units, and even preparation methods. (Formulae in DCMF are derived from more than 700 ancient books with different weight units.) A typical example of ingredients can be listed below: "Ginseng 1 liang, Largehead Atractylodes Rhizome 1 liang, Tuckahoe (peeling) 1 liang, Liquorice Root 8 qian." Identifying such problems of representation granularity and treating them is indispensable for data analysis and decision support.

4.2.2 Representation Consistency

Representation consistency means that data are continuously presented in the same format, consistently represented, consistently formatted, and compatible with previous data [3]. Due to historical reasons, TCM experts thousands of years ago tended to use different notions and expressions to describe one concept. When the information in different dynasties is collected in databases, the problem of representation inconsistency is also introduced. For instance, *ginseng Panax* and *Radix ginseng* could both refer to the Chinese herbal medicine ginseng in English. This situation becomes more complex in Chinese: there are 10 aliases for ginseng. Another example is the inconsistent weight units used in different Chinese medical formulae, which has been mentioned before. Before data analysis and knowledge discovery can be carried out on these data, such representation consistency issues must be addressed to ensure the final reliability.

4.2.3 Completeness

One of the biggest problems hampering the effective usage of TCM resources is the incompleteness of data. Take DCMF for an example; two crucial attributes of DCMF are ingredients and efficacy. The attribute ingredients have already been described, and efficacy is a textual attribute containing a description of the remedy principle in the TCM background. Due to historical reasons, among 85,917 valid records wherein the attribute value of ingredients is not null, only 15,671 records are stored with efficacy not null. That is to say, 81.76% of data in attribute efficacy is missing. Identifying such phenomenon in TCM data and treating this problem is an important task in data analysis.

4.3 Methods to Handle Data Quality Problems

Due to the existence of data quality problems mentioned previously, it is extremely important to conduct necessary data preprocessing activities for data analysis and knowledge discovery. Jiang et al. [19] indicated that data preprocessing was the key to the knowledge discovery of the compatibility rule of TCM formulae. Thus, it is of vital necessity to explore preprocessing methods of TCM data. The data quality problems mentioned in the last section are the main obstacles in TCM on the way to high data quality. In this section, we introduce the preprocessing methods used to handle these problems.

4.3.1 Handling Representation Granularity

The procedure we conduct to treat the representation granularity problem is called structurizing, i.e., to structurize a data field with multiple data elements into multiple separate data fields. To handle the example problem of representation granularity mentioned in the previous section, a concept of a herb information unit (HIU) is defined, which is the name of Chinese herbal medicine, followed by the preparation method, dosage, and weight unit. With this perspective, we could see that the attribute ingredients usually consist of multiple HIUs separated by commas. To effectively use all information in this field, we should first split ingredients into multiple

HIUs. Secondly, for each CIU, we further divide it into four fields: the name of Chinese herbal medicine, preparation method, dosage, and weight unit. To perform this two-step extraction, there are a lot of details and exceptions that should be noticed in practice. For instance, in many records, the delimiter comma might be replaced by semicolon/period, or even be missing; the preparation method/dosage/weight unit is also missing or misspelled in many records. To implement the two-step splitting, a splitting-rule-based system named field splitter was developed in 2003 to handle this problem. Tens of specific splitting rules, such as "keep between A and B" and "replace A with B", are defined. Users can form their own splitting setting by organizing these rules. The system field splitter is found to work well for these years. This is the structurizing method we use to fight representation granularity problems in TCM.

4.3.2 Handling Representation Consistency

To tackle problems of representation inconsistency, especially those in textual/categorical attributes, the basic idea is to refer to the standardized terminology. In TCM context, this task could be done with the help of unified traditional Chinese medical language system (UTCMLS) [12], which is the largest ontology-based language system in TCM. The basic procedure we apply is illustrated in Figure 4.1.

As Figure 4.1 shows, the procedure to handle representation inconsistency consists of three steps. In Step 1, the aliases of all concepts and instances are extracted from UTCMLS, forming a synonym dictionary. In Step 2, by scanning this dictionary from beginning to end, all of the aliases found in target data are replaced with their formal names. In Step 3, a manual check is conducted by the TCM experts to identify incorrect substitutions and new words/phrases that do not exist in UTCMLS. For each new word/phrase, a verification is conducted by experts to see whether it is a true new TCM term or is just a wrongly written version of some term in UTCMLS. In the former case, this new term would be added into the synonym dictionary (or even UTCMLS after a period of time), so that the system could recognize this new term in the next round of scanning. In the latter case, the wrongly written version would be corrected to its right term, and such correcting operation with its context would be automatically recorded as a correction rule by the system. These newly collected correction rules could be automatically applied or manually selected in following sessions. After the whole dataset is treated with these three steps, the synonym dictionary is updated and some correction rules are collected. Then, we could go back to Step 2, performing scanning and substituting, and then to Step 3, checking and updating. After several rounds of processing like these, the problem of representation inconsistency in textual/categorical attributes can be greatly relieved.

Another type of representation inconsistency is the format inconsistency of numerical attributes. For such a type of inconsistency, the solution is to normalize. Let us take the weight unit problem mentioned in the last section as an example. In order to compare the dosages of certain Chinese herbal medicines among multiple formulae, the weight units should be normalized to metric unit "g," according to

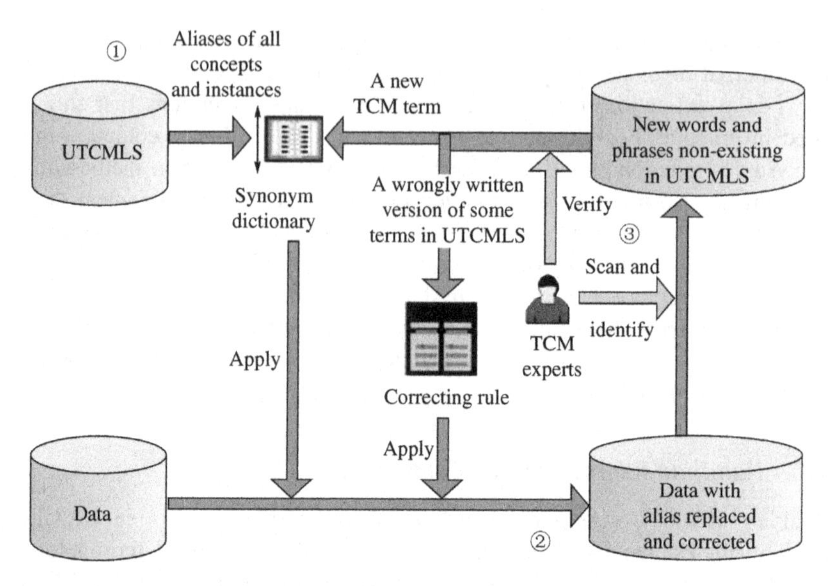

Figure 4.1 Procedure to handle representation inconsistency.

the corresponding conversion rules in different dynasties. Such treatment could be seen as a process of data transformation.

4.3.3 Handling Completeness

Typical methods of a missing value problem include mean imputation, mode imputation, and imputation by regression. However, these approaches are more applicable to structured data than unstructured text. In TCM data, we find most incompleteness problems come from textual attributes. To attack this problem, we propose closest-fit approaches [13] to filling in textual missing values. In closest-fit methods, we search the training set, compare each case with the one having missing value, calculate the similarity between them, and select the case with the largest similarity. Then, we substitute the attribute value of that case for the missing attribute value. Two versions of closest-fit methods could be applied. The first is to search the best candidate in the entire dataset, which can be named global closest fit. The second is to restrict the search space to all cases belonging to the same category, which can be named concept closest fit or category closest fit.

A fundamental issue in this approach is how to evaluate the "closeness" of two given texts. The problem of text comparison is closely related to research areas like string matching, information retrieval, and text mining. To incorporate text similarities in these different disciplines into one framework, we present an order perspective of text similarities. Under this perspective, we propose a new order: semi-sensitive similarity, M-similarity [13], to adapt to missing value application. This similarity is based on three factors: (1) single item matching, (2) maximal sequence matching, and (3) potential matching. Besides, three adjustable parameters are included in M-similarity to make it more flexible. Experiments on real-world TCM datasets show that the meaningful percentage of the concept closest fit method with M-similarity is about 85%, and this number reaches as high as 91.15% for category closest fit. In practice, we also find this closest-fit method can help us a lot to attack incompleteness problems.

4.4 Conclusions

Data quality is a key issue in medical informatics and bioinformatics. In this chapter, we discuss this issue in a subarea of a medical field, i.e., TCM. Due to the high domain specificity of data quality, it is of great necessity to identify key dimensions of data quality in TCM. In this chapter, three data quality aspects are highlighted: representation granularity, representation consistency, and completeness. Moreover, methods and techniques to handle these data quality problems also are presented in this chapter.

Although this chapter is focused on these three dimensions, it does not mean their superior importance over other aspects. They are highlighted because most data quality problems found in TCM practice in past years could fall into these dimensions. By introducing these key aspects and presenting methods to handle the data quality problems, this chapter is expected to help doctors and researchers acquire data with satisfactory quality for more reliable analysis and decision making. The best way to handle data quality problems, whatever area it is, is to avoid them. However, total avoidance is a nearly impossible task in many cases. A realistic and workable strategy is to take data quality into account throughout the process of data collection, storage, and usage, which could help to diminish such problems to the minimum.

References

[1] Ballou DP, Pazer HL. Cost/quality tradeoffs for control procedures in information systems. OMEGA Int J Manage Sci 1987;15(6):509–21.

[2] Wang RY, Reddy MP, Kon HB. Toward quality data: an attribute-based approach. Decis Support Syst 1995;13(3–4):349–72.

[3] Wang RY, Strong DM. Beyond accuracy: what data quality means to data consumers. J Manage Inform Syst 1996;12(4):5–34.

[4] Chengalur-Smith IN, Neely MP, Tribunella T. The information quality of databases. Encyclopedia of database technologies and applications. Idea Group; 2005. p. 281–5.

[5] Kulikowski JL. Data quality assessment. Encyclopedia of database technologies and applications. Idea Group; 2005. p. 116–20.

[6] Danielle GT, Nicolette F. Defining and improving data quality in medical registries: a literature review, case study, and generic framework. J Am Med Inform Assoc 2002; 9(6):600–11.

[7] Rigby M. Integrated record keeping as an essential aspect of a primary care led health service. Br Med J 1998;317:579–82.
[8] Brown PJB, Warmington V. Data quality probes—exploiting and improving the quality of electronic patient record data and patient care. Int J Med Inform 2002;68(1):91–8.
[9] Mikkelsen G, Aasly J. Consequences of impaired data quality on information retrieval in electronic patient records. Int J Med Inform 2005;74(5):387–94.
[10] Feng Y, Wu ZH, Zhou XZ, et al. Knowledge discovery in traditional Chinese medicine: state of the art and perspectives. Artif Intell Med 2006;38(3):219–36.
[11] Zhou XZ, Liu BY, Wu ZH, et al. Integrative mining TCM literature and MEDLINE for functional gene networks. Artif Intell Med 2007;41(2):2007.
[12] Zhou XZ, Wu ZH, Yin AN, et al. Ontology development for unified traditional Chinese medical language system. Artif Intell Med 2004;32(1):15–27.
[13] Feng Y, Wu ZH, Zhou ZM. Combining an order-semisensitive text similarity and closest fit approach to textual missing values in knowledge discovery. In: Proceedings of KES 2005; 2005. LNCS 3682: p. 943–9.
[14] He QF, Cui M, Wu ZH, et al. Compatibility knowledge discovery in Chinese medical formulae. Chin J Inf Tradit Chin Med 2004;11(7):655–8.
[15] Yu T, Jiang XH, Feng Y. Semantic graph mining for e-Science. In: AAAI 2007s workshop on Semantic e-Science (SeS2007); 2007. p. 77–80.
[16] Feng Y, Wu ZH, Zhou ZM. Enhancing reliability throughout knowledge discovery process. In: ICDM'06 workshop of reliability issues in knowledge discovery (RIKD 06); 2006. p. 754–8.
[17] Chen HJ, Mao YX, Zheng XQ, et al. Towards semantic e-Science for traditional Chinese medicine. BMC Bioinform 2007;8(Suppl. 3):S6.
[18] Schmid J. The main steps to data quality. In: Proceedings of the 4th industrial conferences on data mining; 2004. p. 69–77.
[19] Jiang YG, Li RS, Li L, et al. Experiment on data mining in compatibility law of spleen–stomach prescriptions in TCM. World Sci Technol Modern Tradit Chin Med Mater Med 2003;5(2):33–7.

5 Service-Oriented Data Mining in Traditional Chinese Medicine

5.1 Introduction

Problems in the real-world domain are greatly challenging traditional knowledge discovery in a database (KDD) model. As an emerging paradigm of data mining (DM) [1], domain-driven data mining (D^3M) [2] aims to synthesize various intelligence resources in order to support actionable knowledge discovery driven by application problems [2–4]. D^3M has been successfully applied to a variety of applications, as an effective tool to transform data resources into business insight [4]. In particular, it is intended to be used in solution of complex domain problems involving interdisciplinary and cross-organizational collaboration. The collaborative D^3M platform has the following three critical requirements:

1. Share and integrate reusable heterogeneous intelligence resources within a virtual process control environment;
2. Support rich interaction operations for communication between task process and domain experts;
3. Provide efficient data processing methods for specific time costing and storage costing tasks.

It is difficult for existing traditional data mining tools to meet all these requirements because of their limitations on component organization. We need an "intelligence metasynthesis" [3] platform that is resilient and flexible enough to support collaborative D^3M on specific domain problems. According to these requirements, we present the DartSpora, a unified D3M platform that provides data mining as a service. Based on a service-oriented architecture (SOA) [5], DartSpora provides convenient operations on intelligence resources via resource orchestration and service orchestration. The DartSpora system facilitates domain experts to perform knowledge discovery experiments in an intuitive manner. Users can create an experiment by specifying a knowledge discovery process as a service operator tree with customizable properties, executing the process and reviewing the visualized results. Our DartSpora can benefit users with the following technical features:

- *Intelligence metasynthesis*: The platform integrates all kinds of intelligence resources, such as domain knowledge bases, data sources (databases, files in different formats), and agents that contain corresponding functions for serendipitous knowledge discovery and reuse.

Modern Computational Approaches To Traditional Chinese Medicine. DOI: http://dx.doi.org/10.1016/B978-0-12-398510-1.00005-4

- *Resource orchestration for complex D^3M tasks*: The platform is based on a multilayer architecture, containing resource layer, component layer, service layer, and application layer. At the bottom is the resource layer, which contains basic elements such as data resources, algorithms, and communication modules. Above the resource layer is the component layer that is responsible for resource management and task execution. In this layer, a modeled task process is formulated with resource orchestration. The service layer serves as a bridge between the infrastructure and the application layer, by encapsulating the interaction operations and task processes into web services. At the top is the application layer, in which users design D^3M applications and execute them through service assembling.
- *High-performance data mining*: Our platform provides high-performance data mining based on cloud computing through uniformed web services. Taking the use of cloud computing resources as an example, we provide parallel distributed algorithms for a single process that is too large to run on a single machine. These algorithms and their executing environment are integrated as resources of DartSpora platform. As a result, they can be used in the data mining tasks as well as other resources.

The rest of the chapter is organized as follows. Section 5.2 introduces the related work. Section 5.3 describes the system architecture, data mining service operating mechanism, and interaction interface. In Section 5.4, we present a series of case studies in the traditional Chinese medicine (TCM) domain, which proves the effectiveness of the DartSpora platform. Concluding remarks are given in the final section.

5.2 Related Work

Much work has been done with related data mining software and platforms. In our view, this can be classified into three categories: traditional data mining software, data mining system for a specific domain, and a distributed data mining platform. Because DartSpora is developed from a demo (called Spora) that was developed in 2007, we will also give a brief introduction to Spora.

5.2.1 Traditional Data Mining Software

Data mining software can be found in software lists of kdnuggets 2, either commercial software or open-source software.

Commercial software has occupied most of the data mining market. The leading ones such as SPSS Clementine and SAS Enterprise Miner share several common features: graph layout work-flow control, multiple data format support, and enterprise data mining solutions. However, most of the commercial software lacks the extensibility for the user to integrate new algorithms or to modify the existing algorithms for practical needs.

Open-source data mining software also plays an important role in the application of data mining. For example, Weka is a collection of machine learning algorithms for data mining tasks [6] and has contributed a lot to both academic research and applications. RapidMiner, a cross-platform data mining prototyping tool with graphic user interface (GUI), excels in its tree-structured process flow and good extensibility [7]. They provided plenty of algorithms covering most of the data mining methods. Besides, they also introduced an effective process control mechanism. But support of general distributed resources and domain-driven agents is lacking in these open-source tools.

For extensibility and codes of commercial software that are not available, DartSpora gets experiences from open-source projects. By integrating open-source algorithm libraries and using a tree structure put forward by RapidMiner, it can accomplish complex tasks. The platform acts as application service provider to avoid problems such as environment maintenance and expansion modules' version control.

5.2.2 Data Mining Systems for Specific Field

Data mining technology has proved itself valuable when being applied to different domains. But experts in specific domains may not possess professional skills for programming. They need to adapt data mining software to accomplish this work. TCMiner, for instance, is a TCM-oriented mining system [8]. These tools introduced data mining into the application of different areas and have provided valuable references to domain experts. They contribute domain-concerned algorithms and a fixed data mining process to users. But their weakness in extensibility and process management limits their role. It cannot meet requirements such as flexible process formulating and a complex task with multiple resources. DartSpora uses different approaches to solve these requirements. Instead of developing a closed system, we adopt existing achievements to make the platform scalable and easy to expand. Even the distributed parallel algorithms on cloud computing cluster can be added to the task as a phase.

5.2.3 Distributed Data Mining Platform

Data mining tasks differ in resource requirements. Some of them are computation intensive, while some are I/O intensive. Due to the constraints of computational and storage capacity on a single machine, distributed systems for large-scale data mining were developed to satisfy the ever-growing requirements. The most frequently used technology is Grid computing. With the help of Globus toolkit, Grid Miner introduced OGSA-DAI Grid services to gain access to Grid databases [9]. Weka, the popular platform of data mining and machine learning, also has been extended into several Grid-based platforms. For example, Grid Weka enabled the use of Weka in an ad hoc Grid, to separate the entire data mining process into several stages and distribute them among a set of Weka servers. Another Grid-based software derived from Weka is Weka4WS, focused on offering service-oriented distributed data mining in a Grid environment [10].

5.2.4 The Spora Demo

The Spora Demo is a web application with specific data mining methods for TCM. Its purpose is to provide TCM data mining functions through the web. It put forward some new ideas: using RapidMiner's process description and controlled mechanism in TCM experiments, publishing the application through the web. But it lacks support of distributed resources. All available ones must be on the web application server. The structure of Spora Demo is relatively simple, and it is hard for this platform to bear flexible domain-concerned requirements. A reasonable architecture is needed to provide D^{3}M service based on multiple resources.

5.3 System Architecture and Data Mining Service

We introduce DartSpora platform in three aspects: hierarchical structure, service operator organization mechanism, and user interaction. The platform constructs a process control environment based on multiple resources, and it encapsulates the process transaction with SOA to provide flexible data mining services.

5.3.1 Hierarchical Structure

The hierarchical structure assembles the resources and models with the purpose to build up a flexible virtual environment for the data mining process. As shown in Figure 5.1, there are four layers from bottom to top: resource layer, component

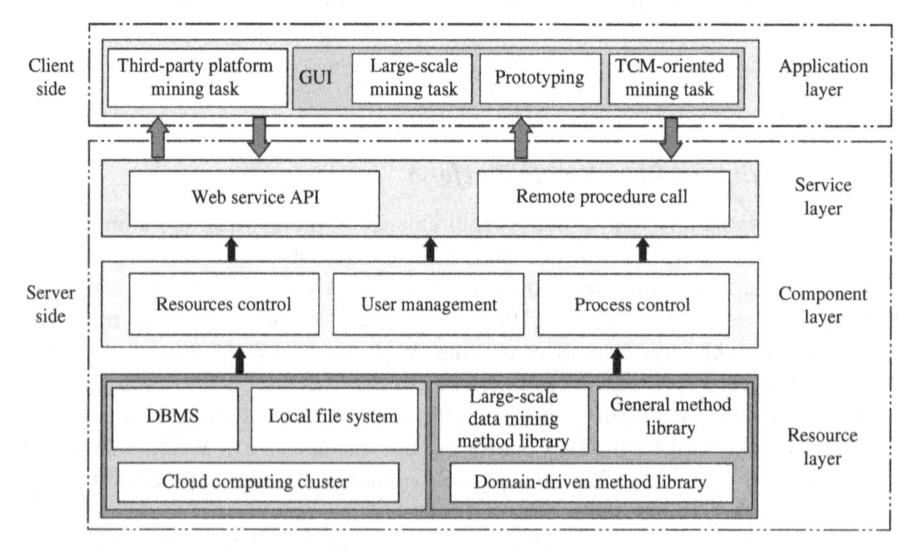

Figure 5.1 Architecture of DartSpora.

layer, service layer, and application layer. The first three layers constitute the main part of the server side, while the last one is assigned to the client side. On the top level of the structure, users can divide the data mining task requirements into subproblems and solve them with a configurable process.

5.3.1.1 Resource Layer

Resource layer is the base of the whole platform. The main parts of this layer are storage resources, computing resources, and algorithm libraries. DartSpora supports various storage and computing resources: database, local file system, cloud computing cluster, and so on. DartSpora uses specific resource units to integrate them; these units control resources by APIs released by the resource server side. But users can ignore these details, for they are hidden under the component layers. The other kinds of resources are data mining methods. We maintained a series of algorithm libraries in this layer. Open-source data mining tools like RapidMiner and Weka have many useful classical algorithm implementations, so we import them for the purpose of general data mining. To deal with domain-concerned data mining problems in the D^3M way, there is also a domain-driven method library, in which algorithms are refined with more domain factors under consideration. This important library is easy to expand in cooperation between data mining experts and domain experts. Besides, distributed parallel algorithms are added with their execution environments together. The method resources are encapsulated into service operators, which is the unit of data mining process' phases.

5.3.1.2 Component Layer

The component layer is an internal bridge between resources and web service. To avoid frequently mapping every part of the resource component into the web service, we designed three components in this layer to control and use resources. The resources control component acts as data interface that generates a data stream for task process. Process control component manages the details of the process: algorithm selection, service-flow design, and parameters configuration in each phrase. All this information is stored in and interpreted from a description file. User Management Module takes charge of users' information management and resource control of the platform. Components of DartSpora are loosely coupled so that external resources can be easily integrated without change of platform architecture.

5.3.1.3 Service Layer

The service layer provides two kinds of web services: one for data mining process configuration services, while the other is for KDD service providing. In the development of user interactions, functions like resources management and data mining process operations GWT-RPC (Google Web toolkit remote procedure call) are used to build up communication between GUI and the server side. It made the construction of the complex interaction mechanism more systematic. The data mining services web APIs for a third-party platform are developed based on SOA. They encapsulated the configured data mining processes into the data mining service. This kind of service can be easily integrated and used by other systems, which, as a result, maximizes the utilization of existing components.

5.3.1.4 Application Layer

Supported by different kinds of web services, applications in this layer can be divided into two categories: domain-driven KDD task design through GUI and a data mining service as an external function through web APIs. On the GUI of this service platform, domain experts can conduct a wide range of data mining applications: data mining prototype experiments, TCM-oriented mining tasks, and large-scale mining tasks with multiple resources involved. Data mining processes mentioned above can be used by a third-party platform as services by calling corresponding web APIs.

5.3.2 Service Operator Organization

As mentioned in Section 5.2, RapidMiner offered an excellent way to describe work flow of a data mining task. It presented process in an operator tree, in which the operator is a unit module for phase control. We extended this method for the service flow presented in DartSpora. Figure 5.2 shows the life cycle of service flow in counterclockwise order. It starts with a service request from application and then a new service process is created and configured. A complete process is composed of service operators that contain various registered resources, and its final pattern can be seen as a service operator tree. It will be executed in depth-first order. When it is finished, the process control module returns the

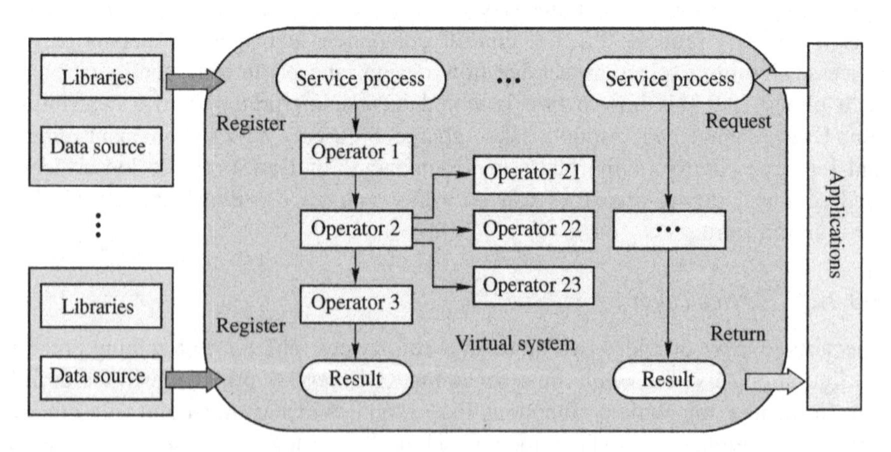

Figure 5.2 The process flow of DartSpora.

result to the client side of the application layer, which is the end of a complete service flow. The virtual environment is component layer based on RapidMiner's experimental work flow mechanism. The process control module collaborated with the resources control module to serve each service process in a single operating environment. Its duties are:

- *To register resources in service operators*: In DartSpora, we use the service operator as the basic unit of a process. It works as a subtask handler. The resources should be registered into a corresponding service operator from externally before it can be used by the process control module. Details of the register procedure are controlled by parameters, and these resources are invoked just before the execution of the whole process.
- *To analyze and execute the service process*: Take the tree in Figure 5.2 as an example, the second phase is actually a subservice process that is composed of three service operators. The tree decides not just the sequence of the steps but also data communication between them. Usually, a service operator only communicates with the predecessor, successor, and parent. The virtual environment initializes service process daemons for each task. These daemons work on the same principle as RapidMiner's experiments. However, the difference is that the objects to be executed are service operators, which make the formulation and execution of the D^3M process possible.

The service configuration procedure is well supported by the user interaction functions of DartSpora. The steps of resources register, service tree construction, and process execution can be graphically displayed in the user interface of the service platform.

5.3.3 User Interaction and Visualization

In D^3M tasks, human intelligence is treated as an important resource, so adequate interaction between humans and the platform is necessary. DartSpora offers a GUI that is presented to the user through the web. The GUI is built up with a variety of widgets to implement operations like the task management, parameter configuration, data view, and program execution.

Figure 5.3 is an overview of DartSpora's interface. The main framework consists of three parts: the task process control region (on the left), the work region (the middle one with most panels), and the wizard region (small region on the higher layer).

The task process control region contains the list of existing experiments and the workspace for the tree structure adjusting of the task process, while the detailed work is accomplished in panels of the work region. For system architecture overview, service operators' access, parameter editing, result displaying, and user management, we prepared corresponding panels. Wizard region is also an important component, which assists domain experts and common users to make specific parameter editing, e.g., file selecting, database connecting, and querying. DartSpora provided a visualization module for the graphical displaying of some nonintuitive text results. A representative case is to state the relationship diseases and Chinese herbs using a network graphic, as shown in the windows in the lower right corner of Figure 5.3.

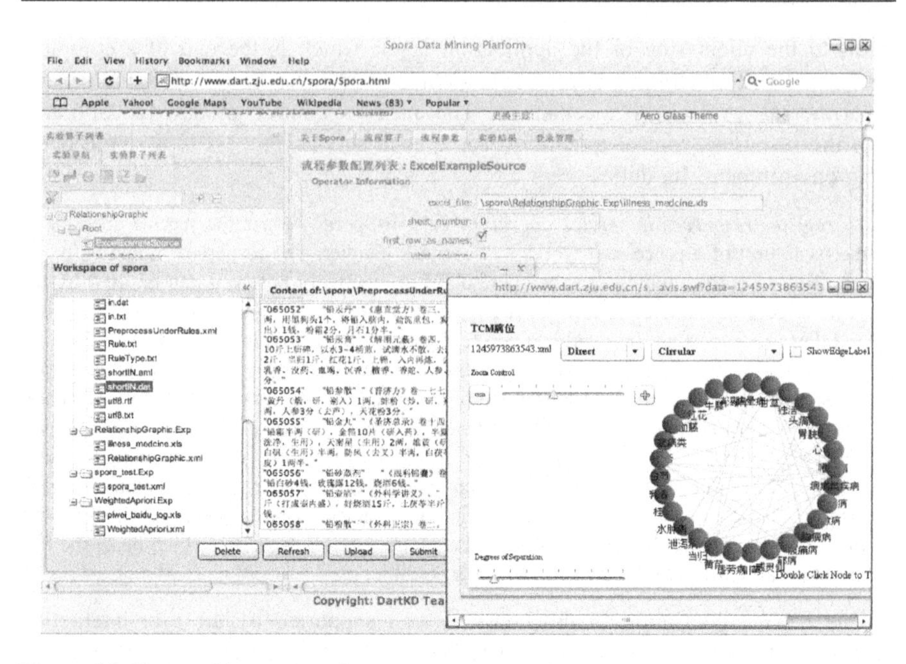

Figure 5.3 The graphic user interface.

5.4 Case Studies

In this section, we use several cases to present the main features of DartSpora. These applications are designed and executed in cooperation with the Traditional Chinese Medicine Institute.

5.4.1 Case 1: Domain-Driven KDD Support for TCM

We developed a series of domain-concerned KDD service operators with the purpose of in-depth pattern mining of TCM prescription data. These service operators can be used to design quality data processing and KDD experiments from scratch, mixed with general ones. Our domain-concerned service operators mainly focus on data preprocessing and in-depth related ingredients pattern mining.

Because TCM prescriptions are written in Classical Chinese which is complex and compact, they should be preprocessed first. However, general data preprocessing methods are not appropriate here, for these data were accumulated over thousands of years, so that the naming and ingredient measurements of prescriptions are ongoing within various criteria. Besides, the contents of prescriptions even describe the process of refining, which acts as noise in ingredients pattern analysis. Our solution is to use transforming rules to describe the exact action. Rules are formulated by domain experts who know well the content of prescriptions. With flexible rules, we can get various objective data from the original data.

The traditional association rule generating methods are established on the assumption that all the input data are credible and each transaction contributes equally to the increase of support value. But de facto TCM data come from ancient medical manuscripts whose authority and quality are different from each other. The prescriptions contained in them should be treated with suitable attention. To quantitate the difference in authority, we add a weight attribute in the itemset generating phase. Different from traditional association rule generating methods such as Apriori [11], each appearance of an itemset contribution is directly proportional to the weight of the corresponding transaction where it is contained. The support$_{\text{weight}}$'s calculation formula is

$$\text{Support}_{\text{weight}}(X) = \frac{\sum_{i \in T_X} \text{weight}_i}{\sum_{j \in T} \text{weight}_j} \tag{5.1}$$

in which, T_X is the collection that contains itemset X, while T denotes the total transaction collection. Based on support$_{\text{weight}}$, we can get confidence$_{\text{weight}}$ that leads to more trustworthy rules:

$$\text{Confidence}_{\text{weight}}(X \Rightarrow X) = \frac{\text{support}_{\text{weight}}(X \cup Y)}{\text{support}_{\text{weight}}(X)} \tag{5.2}$$

These specifically developed methods can achieve more convincing results than general ones. Figure 5.4 is an experiment result that tells the differences between using support and weighted support. Supports of the sixth, eighth, and ninth itemset are lower than their successor, and some of them are out of the selected range. However, in experiments using weighted support, they show themselves in the top list. This result just fits the actual situation.

Frequent Iterms	Support	Weighted Support
人参，白术	0. 19253	2. 33813
人参，甘草	0. 17319	2. 12978
甘草，白术	0. 17425	2. 05514
厚朴，甘草	0. 11387	1. 35832
厚朴，生姜	0. 09057	1. 21351
厚朴，白术	0. 08898	1. 11177
木香，白术	0. 09037	1. 07757
厚朴，姜汁	0. 07176	0. 99795
人参，厚朴	0. 07309	0. 95547
木香，甘草	0. 08051	0. 94483

Figure 5.4 Comparison of support and weight support.

5.4.2 Case 2: Data Mining Based on Distributed Resources

Data mining tasks often need a large amount of computing power and storage. DartSpora supports integration of distributed resources to meet these requirements. In this case, we introduce the use of cloud computing resources. Our cloud computing platform is established with Hadoop cluster, HDFS [12], and web services.

Taking frequent pattern mining as an instance, this kind of method has always a high time complexity. It is impossible to accomplish with large-scale data in a stand-alone environment. A feasible way is to design parallel distributed algorithms that rely on a cloud computing cluster. We redesigned the pattern mining procedure in the map/reduce [13] framework. In that process, the work of itemset counting is mapped into subtasks for distributed mappers, and then the semi-results of mappers are aggregated into the final counting result. As the file operations and algorithm work rely on a cluster that is outside the process control environment, we release them as services above the cloud computing resources for remote access. Corresponding service operators manage to use the service in task processes. Usually, data flows between these service operators are large, so we avoid transferring them directly in the service process but use a temp file in the distributed file system instead. The service operators who control the algorithm details always appear along with ones that work on data upload and data view.

The same integrating method is also applicable to get other kinds of distributed resources. It made the kernel part of DartSpora simple and clear, as well as extendable. At the same time, distributed methods can be used flexibly in different tasks to serve the D^3M.

5.4.3 Case 3: Data Mining Process as a Service

Some data mining processes in DartSpora are valuable if they can be reused in different systems. For this requirement, we encapsulated the process control operation into the web service. Users can control the process on DartSpora through their own programs. The data mining process service used in Assistant Chemistry Research System for TCM (ACRST) is a representative example.

ACRST is an integrated information system developed for the research on the relationship between the symptom and TCM drugs. Many TCM databases and related data sets are involved. Figure 5.5 shows the structure and the main processes of ACRST. The assistant system generates user-specific data sets by executing a series of fixed operations on a background database. These process logics are built based on TCM knowledge and experiences. We can embed data mining processes in the analysis process to supply users with valuable clues.

In ACRST, the data mining process service of DartSpora is embedded into its own operating steps. As shown in Figure 5.4, methods supplied by the data mining service platform are an important way to select valuable information from preliminary data. Here is a concrete example of chemistry components analysis. We use maximal frequent pattern mining to assist the extraction of valuable component combinations. The result of frequent mining contains a lot of redundant information.

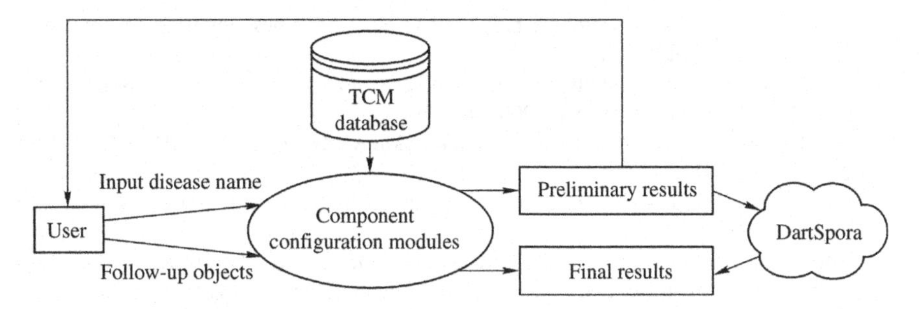

Figure 5.5 The framework of ACRST.

For example, the subpattern of any frequent pattern is also a frequent one. These small patterns are always without any significance in practical analysis. Maximal pattern algorithms only find the frequent patterns that are not a subpattern of others. These longest patterns can reveal active combinations of chemical compositions more effectively.

The data mining process as a service creates a new way to communicate between different resources. The main users of a platform like ACRST are domain experts. They analyzed data on an assistant platform with the help of these services, while their achievements enriched the DartSpora.

5.5 Conclusions

This chapter described a unified D^3M platform DartSpora. The platform is constructed in SOA with multiple layers. We encapsulate a wide range of resources as service operators. These operators could be used to construct various data mining solutions on demand through service orchestration. Processes of these solutions are represented as service operator trees. DartSpora offers data mining solution generation and the solution itself as services. With these services, users of DartSpora could have easy access to the D^3M process supported by multiple resources.

DartSpora is currently used by the Institute of Information on Traditional Chinese Medicine for TCM data research. It cooperates with other data process platforms to provide in-depth data analysis services. We will continuously expand the domain-driven and distributed method libraries. Additionally, we will make more efforts at refinement of the in-depth data mining process and system performance optimization.

References

[1] Han J, Kamber M. Data mining: concepts and techniques. San Francisco, CA: Morgan Kaufmann Publishers; 2006.

[2] Cao L, Zhang C, Liu J. Ontology-based integration of business intelligence. Int J Web Intell Agent Syst 2006;4(4):1–14.
[3] Cao L, Zhang C, Liu j. Intelligence metasynthesis in building business intelligence systems. WImBI 2006, LNAI 4845; 2007. p. 454–70.
[4] Cao L, Zhang C. Domain-driven, actionable knowledge discovery. IEEE Intell Syst 2007;22(4):78–88.
[5] Erl T. Service-oriented architecture: concept, technology, and design. New Jersey: Prentice Hall; 2005.
[6] Holmes G, Donkin A, Witten IH. Weka: a machine learning workbench. Proceedings of the second Australia and New Zealand conference on intelligent information systems; 1994. p. 101–20.
[7] Mierswa I, Wurst M, Klinkenberg R, et al. Yale: rapid prototyping for complex data mining tasks. Proceedings of the 12th ACM SIGKDD international conference on knowledge discovery and data mining; 2006. p. 1–6.
[8] Li C, Tang CJ, Peng J, et al. TCMiner: a high performance data mining system for multi-dimensional data analysis of traditional Chinese medicine prescriptions. In: Wang S, editor. Proceedings ER workshops 2004. Berlin, Heidelberg: Springer-Verlag; 2004. p. 246–57. LNCS 3289
[9] Brezany P, Janciak I, Wőhrer A, et al. GridMiner: a framework for knowledge discovery on the Grid—from a vision to design and implementation. Proceedings of the Cracow Grid Workshop, Cracow, Poland; 2004.
[10] Talia D, Trunfio P, Verta O. Weka4WS: a WSRF enabled weka toolkit for distributed data mining on Grids. Proceedings of PKDD, LNAI 3721, Porto, Portugal, Springer-Verlag; 2005. p. 309–20.
[11] Agrawal R, Srikant R. Fast algorithms for mining association rules in large databases. 20th International conference on very large data bases; 1994. p. 478–99.
[12] Borthaku D. The hadoop distributed file system: architecture and design. Retrieved from: lucene.apache.org/hadoop; 2007.
[13] Dean J, Ghemawat S. MapReduce: simplified data processing on large clusters. Proceedings of OSDI'04: 6th symposium on operating system design and implementation, San Francisco, CA; 2004. p. 137–50.

6 Semantic E-Science for Traditional Chinese Medicine

6.1 Introduction

Traditional Chinese medicine (TCM) is a medical science that reflects traditional Chinese culture and philosophy by formations, factors, and variables to remain in a healthy status. Recent advances in web and information technologies with the increasing decentralization of organizational structures have resulted in a massive amount of TCM information resources (literature, clinical records, experimental data, etc.). Today, a large number of TCM information resources are distributed among many specialized databases like medical formula databases, electronic medical record databases, and clinical medicine databases [1]. For example, the Consortium for Globalization of Chinese Medicine (CG-CM) is a global nonprofit organization, with a mission of advancing the field of Chinese herbal medicine to benefit human kind through joint efforts of academic institutions, industries, and regulatory agencies around the world. Members of CG-CM have TCM information resources of their own. Many TCM scientists and biologists begin to use bioinformatics methods to analyze TCM contents from different points like biochemistry, genetics, and molecular biology, so more and more biology databases have been introduced into TCM research. The information in those databases is potentially related to each other within a TCM knowledge-based system, and it is necessary for TCM scientists to reuse them in a global scope. There is an increasing emphasis on integration of heterogeneous information resources in the presence of such a new setting. TCM scientists need to perform dynamic data integration over hundreds or even thousands upon thousands of geographically distributed, semantically heterogeneous data sources that are subject to different organizations.

Besides the emerging information resources like databases, many scientific methods and processes in TCM have been enclosed as services (e-learning services, information analysis services, data mining services, etc.) by different organizations. There are many bioinformatics services available online, which TCM scientists can use to improve their research from the point of biology. Service-oriented science [2] has the potential to increase individual and collective scientific productivity by making powerful information tools available to all and thus enabling the widespread automation of data analysis and computation. Scientists and applications are able to access web services to finish specific tasks or gain information. As the scale increases, creating, operating, and even accessing services become challenging.

Modern Computational Approaches To Traditional Chinese Medicine. DOI: http://dx.doi.org/10.1016/B978-0-12-398510-1.00006-6

Services are designed to be composed under some contexts, i.e., combined in service workflows to provide functionality that none of the component services could provide alone. There is an increased requirement for coordination of various TCM services to support collaborative and on-demand scientific activities.

E-Science is the term applied to the use of advanced computing technologies to support global collaboration for scientists [3]. However, complete and seamless TCM e-Science is impeded by the heterogeneity and distribution of the independently designed and maintained information and service resources. The use of domain knowledge provides a basis for full interoperability in a distributed environment like the web. Like the foundation of the Semantic Web [4], ontologies [5] are the specification of conceptualizations, used to help programs and humans share knowledge. Encoding domain knowledge in terms of ontologies provides a possible approach to overcoming the problem of semantic heterogeneity of both information and service resources. As mentioned above, there are many information resources in the TCM discipline and most of them exist in terms of databases. Formal semantics in ontologies has provided a feasible way to integrate scientific information resources in a conceptual information space. Besides, some semantic markup languages like OWL-S [6] are used to describe services with more precise semantics. Richer semantics helps to provide automation or semi-automation of activities such as verification, simulation, configuration, composition, and negotiation of services. From this point of view, the research in knowledge-based approaches, especially the semantic techniques, has pointed out a new direction to realize the vision of e-Science for TCM.

A number of approaches for e-Science in biology or medicine have been proposed or developed. Stevens et al. [7] aim to exploit Grid techniques especially the Information Grid to achieve e-Science for bioinformatics. They present the myGrid platform that provides middleware layers to satisfy the needs of bioinformatics. The myGrid platform is building high-level services for data and application integration such as resource discovery, workflow enactment, and distributed query. Tao et al. [8] illustrate through a Semantic Web-based knowledge management approach the potential of applying Semantic Web techniques in an e-Science pilot project called GEODISE for the domain of Engineering Design Search and Optimization. They design advice mechanisms based on semantic matching to consume the semantic information and facilitate service discovery, assembly, and configuration in a problem-solving environment. They have shown the potential of using semantic technologies to manage and reuse knowledge in e-Science. de Roure et al. [9] analyze the state of the art and the research challenges involved in developing the computing infrastructure for e-Science. They proposed the future e-Science research infrastructure, which is termed the Semantic Grid, and a conceptual architecture for the Semantic Grid is presented, which adopts a service-oriented perspective. They consider the requirements of e-Science in the data/computation, information, and knowledge layers. Clearly, e-Science is a wide open research area and there is still much room for improvement in all existing approaches, especially for achieving on-demand e-Science in a knowledge-intensive domain like TCM.

In this chapter, we address the before-mentioned issues by applying semantic techniques and standards such as RDF [10] and OWL [11] to enable database integration and service coordination move toward the full richness of an e-Science vision of TCM science over the Internet. To achieve this vision, we propose an approach (1) to model domain knowledge and develop large-scale domain ontology; (2) to interconnect distributed databases using richer semantics as one huge virtual database; (3) to coordinate scientific services by semantic-driven workflow. We present a dynamic and extendable approach to building on-demand e-Science applications for knowledge-intensive disciplines like TCM based on the semantic techniques. We recognize TCM research as information gathering and process workflows. We have designed and developed the approach as a layered structure to satisfy the TCM research requirements in e-Science. The proposed methods aim at facilitating the integration and reuse of the distributed TCM database and service resources in cyberspace and deliver a semantically superior experience to TCM scientists. We have developed a collection of semantic-based toolkits to facilitate TCM scientists and researchers in information sharing and collaborative research.

6.2 Results

In this chapter, we illustrate our semantic-based approach in a functional perspective as it is developed for different demands of e-Science.

6.2.1 System Architecture

Briefly, we illustrate the abstract architecture of our approach in Figure 6.1. In our approach, a TCM e-Science system is composed of client side and server side. We have designed and developed the server side as a layered structure including resource layer, semantic layer, and function layer.

1. The resource layer mainly supports the typical remote operations of the contents of resources on the web and queries the meta-information of databases and services. The services in this layer extend some core Grid [12,13] services from the Globus [14] platform. We build the whole e-Science system on these Grid services that provide the basic communication and interaction mechanism for TCM e-Science. There are two kinds of services in this layer. Resource Access Service supports the typical remote operations of the contents of databases and execution of services. For relational databases, the operations contain query, insertion, deletion, and modification. Resource Information Service supports inquiring about the meta-information of the database or service resources including relational schema definition, DBMS descriptions, service descriptions, privilege information, and statistics information.
2. The semantic layer is mainly designed for semantic-based information manipulation and integration. This layer is composed of two sublayers. The lower layer contains two kinds of services. Process Semantic Service is used to export services as OWL-S descriptions. Database Semantic Service is used to export the relational schema of databases as RDF/OWL semantic description. The upper layer contains two services. Ontology Service is used to expose the shared TCM ontology and provide basic operations on the ontology.

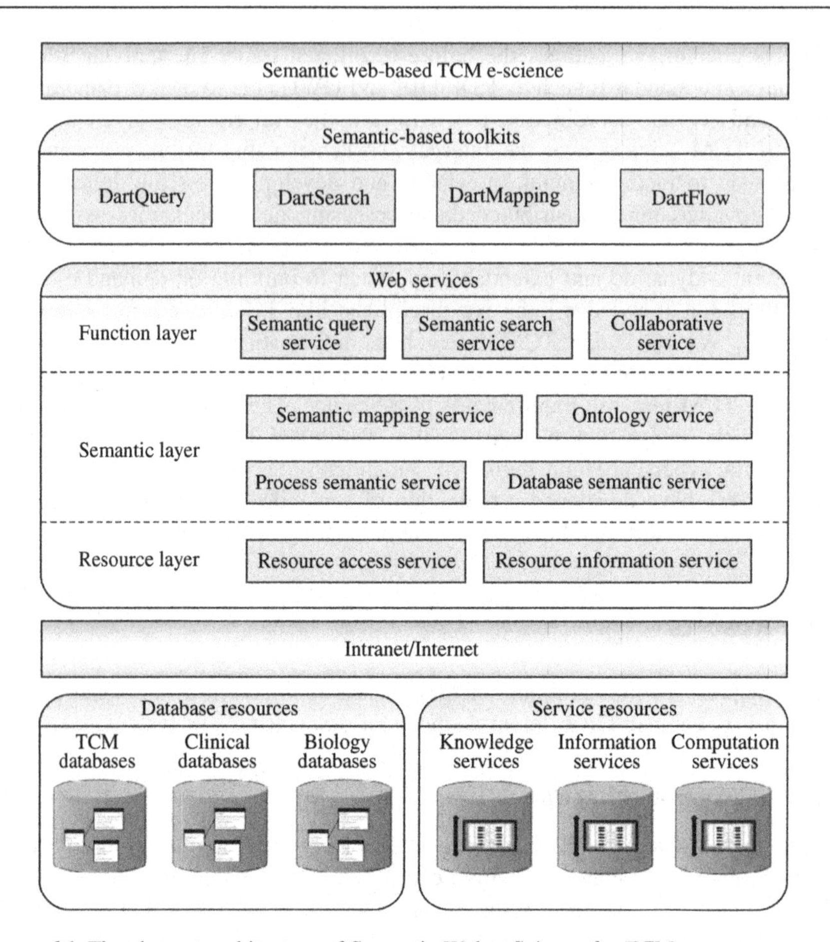

Figure 6.1 The abstract architecture of Semantic Web e-Science for TCM.

Ontology is used to mediate and integrate heterogeneous databases and services on the web. Semantic Mapping Service establishes the mappings from local resources to the mediated ontology. Semantic Mapping Service maintains the mapping information and provides the mechanism of registering and inquiring about the information.

3. The function layer delivers a semantically superior experience to users to support scientific collaborative research and information sharing. Semantic Query Service accepts semantic query, inquires Semantic Mapping Service to determine which databases are capable of providing the answer, and then rewrites the semantic queries in terms of database schema. A semantic query is ultimately converted into a set of SQL queries. The service wraps the results of SQL queries by semantics and returns them as triples. Semantic Search Service indexes all databases that have been mapped to mediated ontology and accepts semantic-based full-text search. The service uses the standard classes and instances from the TCM ontology as the lexicon in establishing indexes. Collaborative Service discovers and coordinates various services in a process workflow to support research activities in a virtual community for TCM scientists.

Note that we differentiate two kinds of services. The services in this architecture are fundamental services to support the whole e-Science system, whereas there are many common services treated as web resources for e-Science process. At the client side, the e-Science system provides a set of semantic-based toolkits to assist scientists to perform complex tasks during research. We call this architecture Dart (Dynamic, Adaptive, RDF-mediated, and Transparent) [15], which is an abstract model for TCM e-Science. A detailed description of the service-oriented architecture is provided in the Methods section.

6.2.2 TCM Domain Ontology

Recent advances in the Semantic Web and bioinformatics have facilitated the incorporation of various large-scale online ontologies in biology and medicine, such as UMLS [16] for integrating biomedical terminology and gene ontology [17] for gene product and MGED Ontology [18] for microarray experiment. As the backbone of the Semantic Web for TCM, a unified and comprehensive TCM domain ontology is also required to support interoperability in TCM e-Science. To overcome the problem of semantic heterogeneity and encode domain knowledge in reusable format, we need an integrated approach to developing and applying a large-scale domain ontology for the TCM discipline. In collaboration with the China Academy of Traditional Chinese Medicine (CATCM), we have taken more than 5 years in building the world's largest TCM domain ontology [19].

We divide the whole TCM domain into several subdomains. The TCM ontology is developed collaboratively by several branches of the CATCM as categories. A category is a relatively independent ontology corresponding to a relatively closed subdomain, compared with the ontology corresponding to the whole domain. There are 12 categories in the current knowledge base of the TCM ontology. Each category is corresponding to a subdomain (Basic Theory of TCM, Formula of Herbal Medicine, Acupuncture, etc.) of TCM. We list the characterization of the content of each category in Table 6.1. Considering medical concepts and their relationships from the perspective of the TCM discipline, we define the knowledge system of the TCM ontology by two components: concept system and semantic system (Figure 6.2). The concept system contains content classes that represent the domain knowledge of the TCM discipline and four kinds of basic implemental classes (name class, definition class, explanation class, and relation class) to define each content class. The semantic system concerns the basic semantic type and semantic relationship of a class. A class has literal property and object property. The range of a literal property is a literal or string, whereas the range of an object property is a class. A content class has five object properties (Table 6.2) with each related to a class. Relation class has two properties: the range of the former is a semantic relationship and the range of the latter is a content class. Content classes are related with each other through semantic relationship. In this way, all content classes in the TCM ontology have the unified knowledge structure, whereas different instances of content class have various contents and relationships.

Table 6.1 TCM Ontology Categories: The Initial Categories Defined in TCM Ontology Corresponding to the Subdomains of TCM

Categories
The Basic Theory of Traditional Chinese Medicine
The Doctrines of Traditional Chinese Medicine and Relevant Science
Chinese Materia Medica (Herbal Medicine)
Formula of Herbal Medicine
Humanities
Medicinal Propagation and Other Resources
Cause and Mechanism of Disease and Diagnosis
Therapeutic Principles and Treatments
Informatics and Philology
Acupuncture
Prevention
Diseases

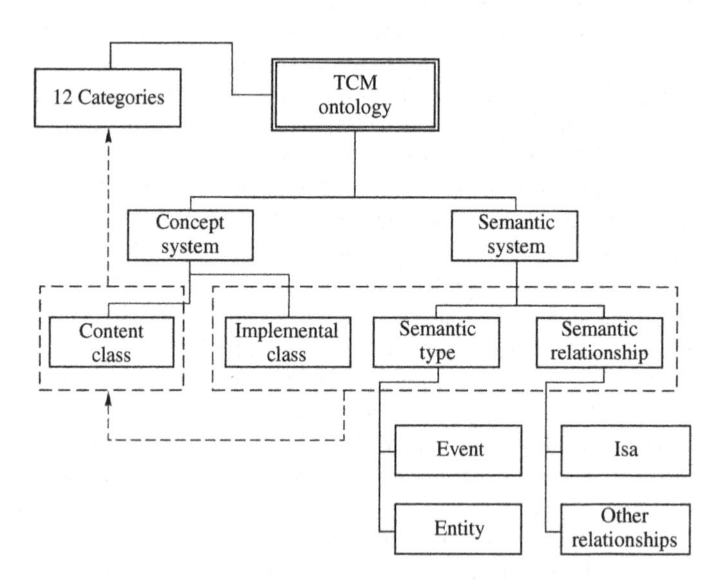

Figure 6.2 The semantic system framework of the TCM ontology.

There are more than 20,000 classes and 100,000 instances defined in the TCM ontology, and the ontology has become a distributed large-scale knowledge base for TCM domain knowledge. The ontology has become large enough to cover different aspects of the TCM discipline and is used to support semantic e-Science for TCM. As a large-scale domain ontology, the TCM ontology is used to integrate various database resources from a semantic view and provide formal semantics to support service coordination in TCM e-Science.

Table 6.2 Content Class structure: The Structure of Content Class in the TCM Ontology

	Class Property	Property Value
hasNames	Name Class	
hasDefinitions	Definition Class	
hasExplanations	Explanation Class	
hasRelatedClasses	Relation Class	Semantic Relationship Content Class
hasSemanticTypes	Semantic Type	

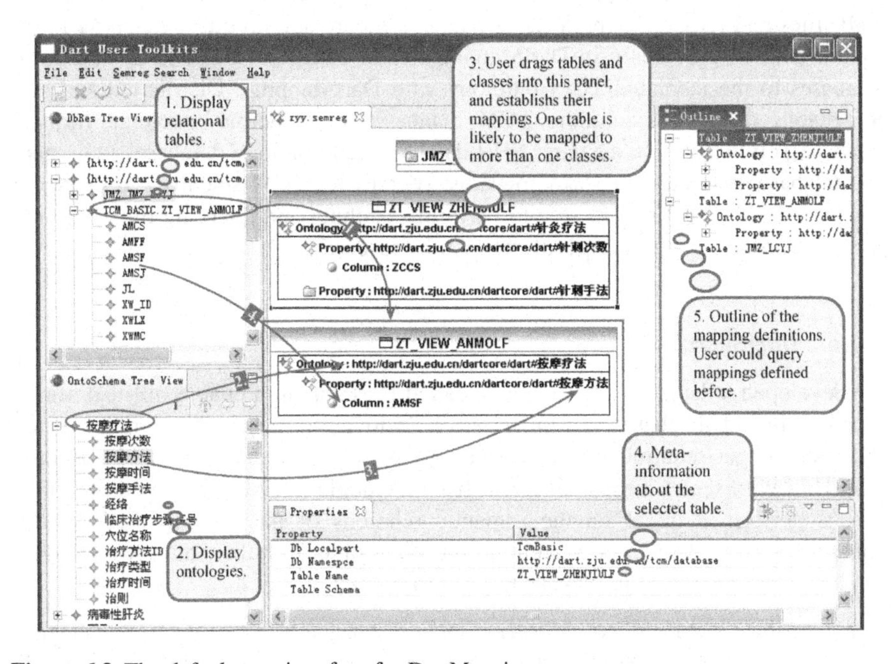

Figure 6.3 The default user interface for DartMapping.

6.2.3 *DartMapping*

In our approach, the before-mentioned domain ontology acts as a semantic mediator for integrating distributed heterogeneous databases. Relational schemata of distributed TCM databases are mapped to the TCM ontology according to their intrinsic relations. To facilitate the process of semantic mapping between the schemata of local databases and the semantics of the mediated ontology, we have developed a visual semantic mapping tool called DartMapping for integrating relational databases in a Semantic Web way (Figure 6.3). The tool provides two major functions: establishing semantic mapping from heterogeneous relational database to a mediated ontology semi-automatically, especially mapping for composite schema with a complex join between tables, and converting relational databases schema to ontology statements based on the semantic mapping information.

Figure 6.3 depicts how we use DartMapping to establish mapping between ontology and database schema. Relational database schema is displayed in the hierarchy including the names of databases, tables, and the corresponding fields (1). The class hierarchy and class properties of the mediated ontology are displayed below (2). Classes and properties are displayed as labels in the panel. Users drag tables and classes into the main panel (3) and establish their mappings directly. One table is likely to be mapped to more than one class. The meta-information about the selected table is shown under the main panel (4). The right panel shows the outline of the mapping definitions (5). A mapping definition can be exported as XML files and reused by applications. Also, users are able to query mapping information defined previously in DartMapping. TCM scientists are able to map local databases to the mediated TCM ontology with DartMapping. Distributed and heterogeneous databases including TCM databases (e.g., herbal medicine formula database), clinical databases (e.g., EHR database), and biology databases (e.g., neuron database) are integrated as knowledge sources for TCM scientists to carry out research. TCM scientists are able to perform searching and querying over the integrated databases to gain useful information by research.

6.2.4 DartSearch

We developed a database search engine called DartSearch to enable full-text search over distributed databases. Scientists are able to perform searching in integrated databases to get required information as we do with search engines like Google [20]. However, search here is different from a Google-like search. The search process is performed based on the semantic relations of the ontology. We call it semantic search, which is searching for data objects rather than web pages. Semantics are presented in two aspects in DartSearch:

1. We construct a domain-specific lexicon for segmentation based on the TCM ontology. Each term in the lexicon is a class or instance in the ontology plus its part of speech. When we segment a piece of information from the database, only the words that appear in the TCM ontology are segmented whereas other words are discarded as irrelevant information to TCM semantic search.
2. Unlike a key word-based index in a traditional search engine, we construct an index for classes or instances in databases. The semantic relations between those classes or instances are encoded as part of the index.

In this way, scientists are able to search with more accurate constraints and get more relevant information from search results. For example, if a TCM scientist wants to find some TCM formulae that cure influenza, then he can use "influenza" as a key word to perform a semantic search. The search returns TCM-specific information and the information that does not contain the key word influenza but contains terms related to influenza is also returned. We connect directly matched information and relevant information by using semantic relations in the ontology.

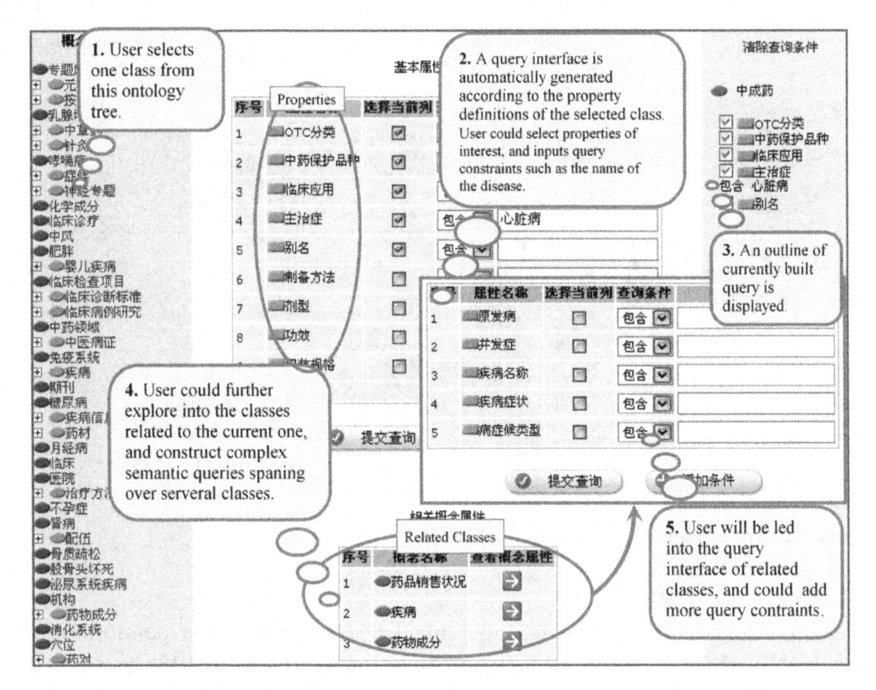

Figure 6.4 The default user interface for DartSearch.

We provide users with a Google-like search interface to perform a semantic search (see the bottom left panel in Figure 6.4) in DartSearch. The result of a semantic search request is shown in Figure 6.4 with "gene" used as the key word. The statistics information about the search result (e.g., the number of items) is displayed (1). DartSearch lists the items in a descending order according to their matching degrees to the key words of the search (2). Each item in the list is a piece of information from databases that have been mapped to TCM ontology classes (3). At the bottom of each item there are the classes the item is mapped to (4), the classes relevant to the mapping classes (5) and the matching degree to the key word. The classes and relevant classes are connected by semantic relations in the ontology. The schemata of a database are allowed to be mapped to several categories of the TCM ontology. Categories that are relevant to the search result are listed in a descending order according to their matching degrees to the search (6).

6.2.5 DartQuery

Generally, semantic search only gives us a coarse set of results. If scientists want to get more exact information, they are able to perform querying instead of searching in the semantic layer. A web-based query tool called DartQuery is provided for scientists to query over distributed TCM databases dynamically (Figure 6.5).

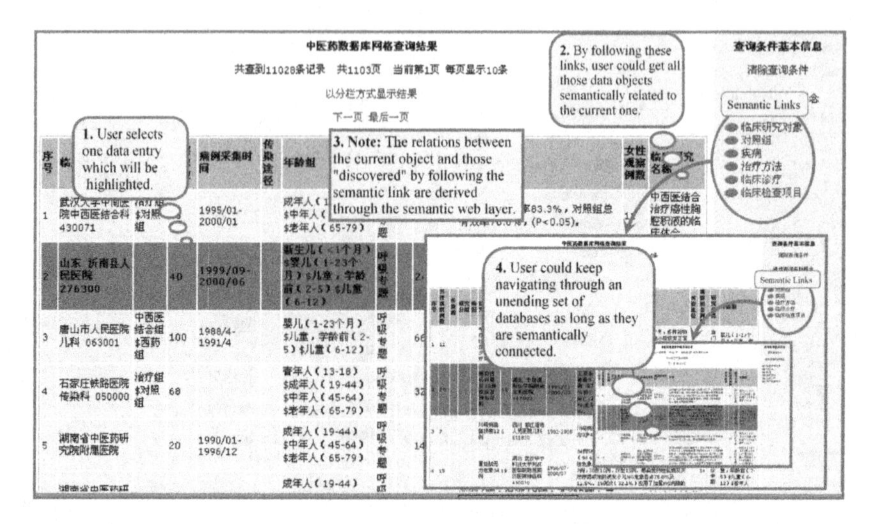

Figure 6.5 The default user interface for DartQuery.

Relevant categories generated during semantic search imply the possible scopes from within which scientists perform semantic query. They are able to select the category with the largest matching degree to construct a semantic query statement. To enable querying in the semantic layer, we use the SPARQL [21] query language. Every query in SPARQL is viewed as an ontology class definition, and processing a query request is reduced as computing out ontology instances satisfying the class definition [22]. The statement of a semantic query about the properties (name, usage, composition, etc.) of a TCM formula that cures influenza is as follows:

```
SELECT?fn?fu?fc?dn?dp?ds
WHERE {
?y1 rdf: type tcm: Formula_of_Herbal_Medicine
?y1 tcm: name?fn
?y1 tcm: usage_and_dosage?fu
?y1 tcm: composition?fc
?y1 tcm: cure?y2
?y2 tcm: name "influenza"
?y2 tcm: pathogenesis?dp
?y2 tcm: symptom_complex?ds
}
```

Such a query in SPARQL is constructed dynamically. A form-like query interface is used to facilitate users in constructing semantic query statements in a web browser. The user interface incorporates an open-source AJAX framework [23], which enables immediate data update without refreshing web pages in web browsers. DartQuery generates querying forms automatically according to the class definitions in a category. Scientists who want to query something are able to construct a query statement by selecting classes and properties from the forms in the query interface. Figure 6.5 depicts the process of how a user constructs a semantic query about traditional patent medicine. Starting from the ontology view panel on the left, users are able to browse the hierarchy tree and select the relevant classes (1). A query form corresponding to the property definitions of the selected class is automatically generated and displayed in the middle. Users could select properties of interest and input query constraints such as the efficiency of the medicine (2). An outline of the currently built query including the current class is displayed (3). Users could further explore into the classes related (e.g., disease) to the current one and construct more complex semantic queries spanning several classes (4). Users are led into the query interface of related classes and could add more query constraints (5).

The SPARQL query statement is submitted to the system and converted into a SQL query plan according to the mapping information between database schemata and the mediated ontology. The SQL query plan is then dispatched to specific databases for information retrieval. The query returns all satisfactory records from databases that have been mapped to the ontology. Because the query result from databases is just a record set without any semantics, the system converts the record set into a data stream in RDF/XML format and the semantics of the result is fully presented. Figure 6.6 depicts the situation in which a user is navigating the query

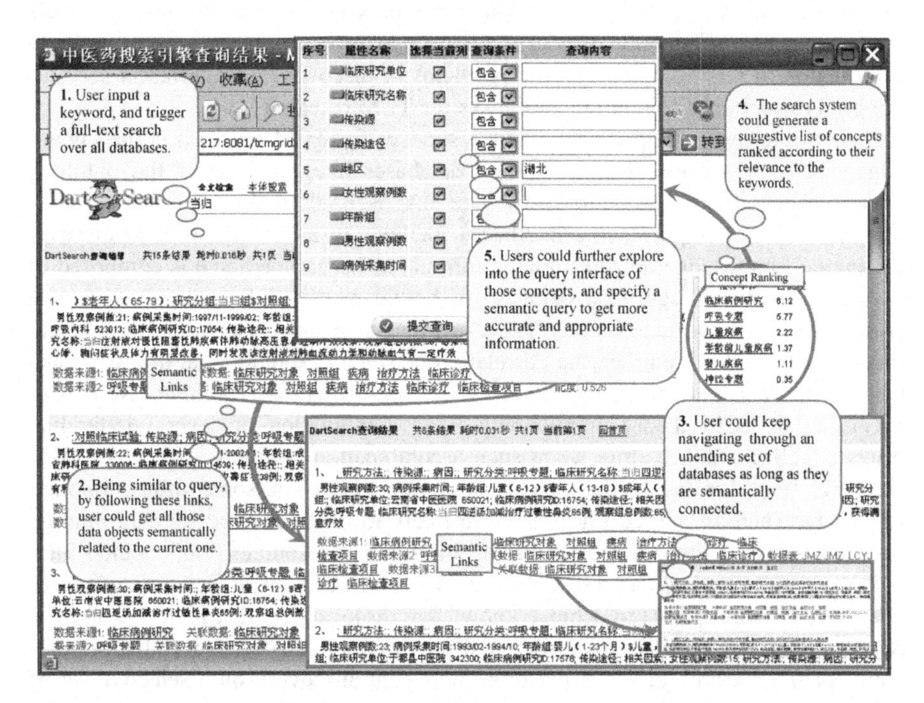

Figure 6.6 The query result of DartQuery.

result. The statistical information about the query result is displayed (1). Users select one data object, which is highlighted (2). By following semantic links, users could get all those data objects semantically related to the current one (3). Note that the relations between the selected object and those discovered by following semantic links are derived from the ontology in the semantic layer. Users could keep on navigating through a collection of databases as long as they are semantically connected (4).

6.2.6 TCM Service Coordination

Ontology semantics are used to support dynamic and on-demand service coordination in a VO. Scientists are able to discover, retrieve, and compose various services to achieve complex research tasks in a visual environment.

6.2.7 Knowledge Discovery Service

There are various services in a TCM VO and we mainly recognize three kinds of services: computation services, information services, and knowledge services. Computation services are services that execute computational jobs or analyze scientific data. Information services are services that manipulate and provide specific information. Semantic query service and semantic search service mentioned before are two typical information services. Knowledge services are services that apply information to solve domain-specific problems or discover facts. Different services are used to support different kinds of tasks for TCM research.

One of the most important knowledge services for TCM research is the knowledge discovery service. The distributed databases integrated under the ontology contain much implicit domain knowledge that is hard to discover manually by humans and thus require some intelligent methods to assist scientists to discover the implicit knowledge. For example, a formula of herbal medicine is composed of several individual drugs. In a database of herbal medicine formulae, we get the components of a formula directly; however, the same individual drug may appear in several formulae, and then the correlation between two individual drugs in various formulas cannot be acquired directly by querying or searching. Note that, according to TCM theory, a relatively fixed combination of several individual drugs is called a paired drug when such a combination is able to strengthen their medical effects or lessen the toxicity and side effects of some drugs. Implicit knowledge such as "paired drug" is more likely to be discovered by data mining, instead of directly querying or searching information resources. Our method integrates several semantic-based data mining algorithms like the associated and correlated pattern mining [24] to achieve knowledge discovery on distributed databases. Scientists are able to select a knowledge discovery service according to the requirements of the research task and perform knowledge discovery on a selective set of information from distributed databases.

6.2.8 DartFlow

Besides database integration, a sophisticated e-Science system should also support service coordination for scientists, which is a significant part of TCM e-Science. Similarly with bioinformatics, TCM scientific research often requires coordination and composition of service resources. We have applied semantic techniques to achieve dynamic and on-demand service coordination in a VO and developed a web-based service coordination tool (Figure 6.7) called DartFlow [25]. DartFlow provides a convenient and efficient way for scientists to collaborate with each other

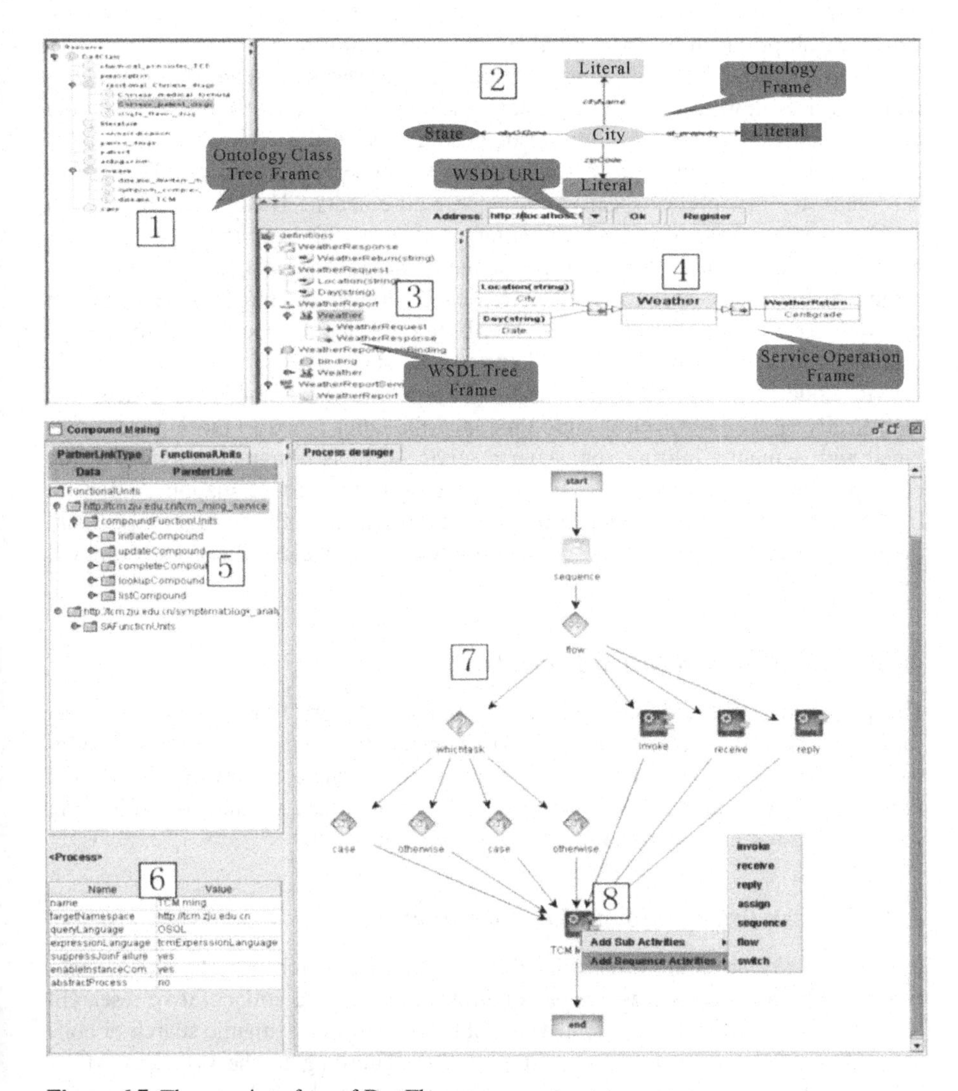

Figure 6.7 The user interface of DartFlow.

in research activities. It offers interfaces to allow researchers to register, query, compose, and execute services in the semantic layer.

Service providers register a component web service into the VO before service composition. DartFlow integrates a service registration portal for scientists to register new services. The class hierarchy (1) and class properties (2) of the mediated ontology are displayed graphically. Service description (e.g., the input and output parameters) is displayed in the hierarchy (3). Similarly with semantic mapping in database integration, service providers create mappings between ontology classes and service descriptions (4). The mapping information is stored in the repository of the portal. Automatic service discovery and service matchmaking is achieved based on semantics. So far, DartFlow has been full of a collection of scientific services, which are all provided by different TCM research institutes.

When a VO has been filled with various applied services, scientists are able to build service flow to achieve complex research tasks in DartFlow. We should retrieve enough services in order to compose a service flow. If scientists want to query services, they submit a service profile (e.g., a service to analyze TCM clinical data) to the portal specifying their requirements. The portal invokes a suitable matchmaking agent to retrieve target services for users (5). The agent has been implemented according to some semantic-based service matchmaking algorithms. Scientists are able to compose retrieved services (6) into a service flow in the workspace (7) to achieve a research task. In order to enhance the flexibility and usability of service flow, DartFlow supports both static activity node and dynamic activity node in service flow (8): the former refers to those nodes combined with specifically applied services at build time; and the latter refers to those nodes combined with semantic information. After a service flow is designed graphically, the corresponding OWL-S file is generated according to the service mapping information. Scientists are able to validate the service flow from both the logic aspect and the syntax aspect with a validator in DartFlow and the validated service flow will be executed ultimately.

6.2.9 TCM Collaborative Research Scenario

The proposed semantic-based approach is able to support TCM scientists to perform research collaboratively in a VO. TCM scientists are able to use the semantic-based toolkits mentioned before in web browsers anywhere to solve problems and finish tasks. We illustrate the application of our approach through the following collaborative research scenario as several steps (Figure 6.8).

6.2.10 Task-Driven Information Allocation

Information resources are often related to perform a research task. Grouping task-related information resources is a precondition for achieving collaborative research. Given a research task, TCM scientists are able to perform a semantic search or construct a semantic query to allocate information according to the task context. A TCM scientist, say Wang, is performing a research task about the impact of herbal

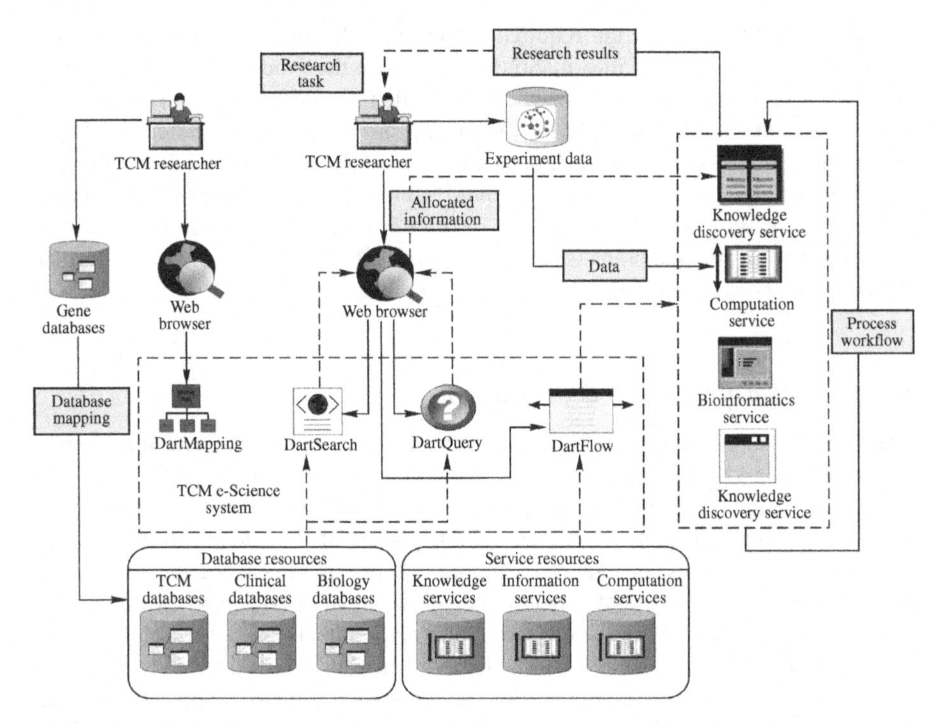

Figure 6.8 A scenario of TCM collaborative research based on the TCM e-Science system.

medicine formula on gene expression. As a TCM scientist, Wang is not familiar with biology, especially genes, so he needs to find some initial information about genes before starting to conduct experiments. He is able to perform a semantic search over distributed databases in DartSearch about genes as well as their relations with TCM. DartSearch will return a general result about genes and their relation to TCM. If Wang wants more exact results of current research progress on herbal medicine formulae and genes, he can perform a semantic query in DartQuery. The semantic search in DartSearch has implied that the required information is mainly located in category "Formula of Herbal Medicine and Diseases." Then Wang is able to perform a semantic query within the databases that have been integrated under these two categories. Wang constructs semantic query statements dynamically in DartQuery and the query returns a collection of literature about herbal medicine formulae and genes.

6.2.11 Collaborative Information Sharing

After reading a batch of relevant papers, Wang decides to perform further research about the relations between herbal medicine formulae and gene expression. However, he finds the information he allocated is insufficient for his task, and it means the TCM VO lacks the required information. Scientists are able to allocate only a very small subset of information or services in the field of TCM. It is impossible for a single scientist to deal with all the domain information. Scientists can share information collaboratively in a VO based on the semantic e-Science system. Wang can communicate with other scientists in the VO to ask for required information. Fortunately, an institute in the VO has developed a new database that contains information about gene expression. The institute registers the database into the VO by creating semantic mappings with DartMapping. Then Wang is able to get further information about gene expression by querying the database.

6.2.12 Scientific Service Coordination

Wang selects suitable services according to his research requirements and designs a service flow in DartFlow to achieve his research goal (see Figure 6.8). The first knowledge discovery service in the service flow is used to discover some underlying rules from the allocated information. The result of knowledge discovery has shown that there exists an underlying relation between Sini decoction (a kind of herbal medicine formula) and glutathione S-transferase (GST) gene expression in many research papers. Wang starts to conduct experiments on the impacts of Sini decoction on GST gene expression. The experiment data is submitted to the computation service in the service flow. He also uses bioinformatics services such as BLAST in the service flow to deal with the works related to the GST gene. The final result of the service flow has shown that Sini decoction has strong impacts on GST gene expression. The service flow here may involve a recursive process in order to refine the result.

6.3 Discussion

Due to the bottleneck of information extraction and NLP, the proposed approach is inclined to structured information resources rather than unstructured or semi-structured resources. However, much information is involved in those resources, which we cannot integrate well into the TCM e-Science system with the current method. Although we could extract schemata from unstructured or semi-structured resources and map those to the mediated ontology in a similar way as database integration, this leaves much work to be done for the purpose. We have provided a set of semantic-based toolkits to assist TCM scientists to reuse information and carry out research. Although the tools are implemented and used based on a web browser, the process of interaction may be still a little bit complex to TCM scientists who have no knowledge of semantics. As TCM is a traditional science, there are also many TCM scientists who are not even familiar with computers and the Internet. For those scientists, we should improve the usage and convenience of the system to satisfy their requirements well.

6.4 Conclusions

We have presented a comprehensive and extendable approach that is able to support on-demand and collaborative e-Science for knowledge-intensive disciplines like TCM based on semantic and knowledge-based techniques. The semantic-based e-Science infrastructure for TCM supports large-scale database integration and service coordination in a virtual organization. We have developed a collection of semantic-based toolkits to facilitate TCM scientists and researchers to achieve information sharing and collaborative research. We illustrate the application of the proposed approach through a TCM collaborative research scenario. Based on the proposed approach, we have built a fundamental e-Science platform for TCM in CATCM and the system currently provides access to over 50 databases and 800 services in practice. The result has shown that integrating databases and coordinating services with a large-scale domain ontology is an efficient approach to achieving on-demand e-Science for TCM and other similar application domains, such as life science and biology.

6.5 Methods

In our research, we use domain ontologies to integrate TCM database resources and services in a semantic cyberspace and deliver a semantically superior experience. Related methods include TCM Ontology Engineering, View-based Semantic Mapping, and Semantic-based Service Mapping.

6.5.1 TCM Ontology Engineering

TCM ontology is a basic element to achieve semantic e-Science for TCM; therefore, the quality of the ontology directly affects the e-Science. We should develop the TCM ontology according to some criteria based on agreement among the participant institutes. The development of ontologies is a modeling process that needs the cooperation of ontology engineers (also called ontologists) who should have sufficient understanding of the domain knowledge. In our experience, ontology construction is a complex and labor-intensive activity.

First, we employ a layered privilege model in ontology development. Users who play different roles in the process of ontology development hold different privileges. There are mainly four kinds of privileges: reader, editor, checker, and administrator.

1. Ontology readers are able to browse all the contents of the ontology.
2. Ontology editors are able to input, modify, and delete instances within a category but have no privilege to manipulate the classes of the category.
3. Ontology checkers own the privilege to manipulate both classes and instances in a category.

4. Ontology administrators have the global privilege to access all categories of the whole ontology. Then, we could develop the ontology according to the following procedure:
   Step 1: Analyze and determine knowledge sources. The scientific control of the conceptual glossary of a discipline is the most important issue in NLP. It is necessary to analyze specialized lexicons as the knowledge sources of ontology contents.
   Step 2: Construct an upper-level conceptual framework. Comprehensive analysis and research of the disciplines is needed before the ontology design. A domain-oriented conceptual framework is constructed to address all the knowledge engineering problems and instruct editing ontology content.
   Step 3: Determine and assign developing tasks. Developing a large-scale ontology is a laborious task that requires collaborative efforts. We divide a large-scale ontology into categories and assign developing tasks to participants by category according to the complexity of each category.
   Step 4: Extend conceptual hierarchy. Checkers create low-level class hierarchy.
   Step 5: Materialize ontology contents. Editors extract and acquire domain knowledge from various sources and formalize knowledge into instances.
   Step 6: Check and revise contents. Checkers check each instance in the category they take charge of to make sure that there is no error or contradiction in newly input contents.
   Step 7: Publish ontology by using user interface. The ontology is published as a web service and users are able to browse and query the ontology through web browsers.

Step 4 to step 6 is a recursive process. Follow this general procedure, we are able to develop a large-scale domain ontology for the e-Science system.

6.5.2 View-Based Semantic Mapping

The semantic-based e-Science system allocates database information and integrates heterogeneous databases together under the TCM ontology by creating semantic views. A relational table is mapped into one or more classes of the ontology and a table field is mapped to a class property. Implicit relationships between database resources are interpreted as semantic relations in the ontology.

According to the conventional data integration literature [26], the view-based approach has a well-understood foundation and has proven to be flexible for heterogeneous data integration. There are two kinds of views in conventional data integration systems, GAV (global-as-view) and LAV (local-as-view). Considering the Semantic Web situation, GAV is to define each class or property as a view over relational tables, and LAV is to define each relational table as a view (or query) over the mediated ontology. The experiences from conventional data integration systems tell us that LAV provides greater extensibility than GAV: the addition of new sources is less likely to require a change to the mediated schema [26]. In the field of TCM, new databases are regularly added so the total number of databases is increasing gradually.

Therefore, the LAV approach is employed in our method, i.e., each relational table is defined as a view over the ontologies. We call such a kind of view a semantic view, and such a kind of mapping from a relational database to ontology as semantic mapping. Like that in conventional data integration, a typical semantic view consists of two parts: the view head, which is a relational prediction, and the view body, which is a set of RDF triples. In general, the view body is viewed as a query of the ontology, and it defines the semantics of the relational prediction from the perspective of the ontology. The meaning of semantic view would be clearer if we construct a target instance based on the semantic mapping specified by these views. In this way, different TCM databases can be integrated under the shared TCM ontology. Scientists need not care about the actual structure of database resources, and they just operate on the semantic layer. More detailed aspects about semantic view and semantic mapping can be found in Ref. [27].

6.5.3 Semantic-Based Service Matchmaking

A service is abstracted as a service description including input and output parameters. If the service descriptions are mapped to ontology classes, service matchmaking and composition can be achieved automatically and dynamically based on semantics. Ideally, given a user's objective and a set of services, an agent would find a collection of service requests that achieve the objective. We use a semantic-based method to achieve dynamic service matchmaking and composition in DartFlow. Assume X and Y are two ontology classes. We represent the matching degree of class X to class Y as Similarity(X, Y). If X can provide all the properties that Y embodies, they are totally matched. If X embodies Y, they are partially matched. They only partially provide the properties that value of Similarity(X, Y) ranges from 0 to 1. The value 0 means X is not semantically similar to Y at all, and the value 1 means X is the same as Y. Note that Similarity(X, Y) and Similarity (Y, X) represent different matching degrees. Different relations between X and Y result in different formulae of similarity evaluation.

Service matchmaking is to match a service request against a collection of services. As services are mapped to ontology classes by service providers, service matchmaking is reduced to calculating the semantic similarity of ontology classes. Given a service $S = (I, O)$ and a service request $R = (I_r, O_r)$, we can calculate the matching degree between S and R, which is denoted as $\Omega(S, R)$. $\Omega(S, R)$ is mainly determined by the similarity between I and I_r and the similarity between O and O_r. Our algorithm ensures the value of a matching degree ranging from 0 to 1. Service composition is also performed based on semantic similarity. Given a service $A = (I_a, O_a)$ in the service flow and a collection of candidate services, the service $B = (I_b, O_b)$ will be selected as the subsequent service of A in the service flow as long as $\Omega(A, B)$ is the largest among the candidate services. More detailed aspects of the algorithm can be found in Ref. [25]. Given a representation of services as actions, we can exploit AI planning techniques for automatic service composition by treating service composition as a planning problem [28].

References

[1] Feng Y, Wu Z, Zhou X, et al. Knowledge discovery in traditional Chinese medicine: state of the art and perspectives. Artif Intell Med 2006;38(3):219–36.

[2] Foster I. Service-oriented science. Science 2005;308(5723):814–7.
[3] de Roure D, Hendler JA. E-Science: the Grid and the Semantic Web. IEEE Intell Syst 2004;19(1):65–71.
[4] Berners-Lee T, Hendler J, Lassila O. The Semantic Web. Sci Am 2001;284(5):34–43.
[5] Gruber T. A translation approach to portable ontology specifications. Knowl Acquis 1993;5(2):199–220.
[6] OWL-S. <http://www.w3.org/Submission/OWL-S/>.
[7] Stevens R, Robinson A, Goble CA. MyGrid: personalised bioinformatics on the information Grid. Bioinformatics 2003;19(Suppl. 1):i302–304.
[8] Tao F, Shadbolt N, Chen L, et al. Semantic Web based content enrichment and knowledge reuse in e-Science. Proceedings of 3rd international conference on ontologies, databases, and applications of semantics for large scale information systems, October 25–29, Agia Napa, Cyprus; 2004. p. 654–69.
[9] de Roure D, Jennings NR, Shadbolt NR. Research agenda for the semantic Grid: a future e-Science infrastructure. http://www.semanticgrid.org/v1.9/semgrid.pdf/>; 2003.
[10] Resource description framework (RDF). <http://www.w3.org/TR/rdf-concepts/>.
[11] Web ontology language (OWL). <http://www.w3.org/TR/owlfeatures/>.
[12] Foster I, Kesselman C, editors. The Grid: blueprint for a new computing infrastructure. San Francisco, CA: Morgan Kaufmann; 1999.
[13] Foster I, Kesselman C, Tuecke S. The anatomy of the Grid: enabling scalable virtual organizations. Lect Notes Comput Sci 2001;2150:1–26.
[14] Foster I, Kesselman C. Globus: a metacomputing infrastructure toolkit. Int J Supercomput Appl 1997;11(2):115–28.
[15] Chen H, Wu Z, Mao Y, et al. DartGrid: a semantic infrastructure for building database Grid applications. Concurr Comput 2006;18(14):1811–28.
[16] Bodenreider O. Unified Medical Language System (UMLS): integrating biomedical terminology. Nucleic Acids Res 2004;32(D):D267–70.
[17] Ashburner M, Ball CA, Blake JA, et al. Gene ontology: tool for the unification of biology. Nat Genet 2000;25:25–29.
[18] Whetzel P, Parkinson H, Causton H, et al. The MGED ontology: a resource for semantics-based description of microarray experiments. Bioinformatics 2006;22 (7):866–73.
[19] Zhou X, Wu Z, Yin A, et al. Ontology development for unified traditional Chinese medical language system. Artif Intell Med 2004;32(1):15–27.
[20] Google. <http://www.google.com/>.
[21] SPARQL query language for RDF. <http://www.w3.org/TR/rdfsparql-query/>.
[22] Chen H, Wang Y, Wang H, et al. Towards a Semantic Web of relational databases: a practical semantic toolkit and an in-use case from traditional Chinese medicine. Proceedings of the 5th international Semantic Web conference, November 5–9, Athens, GA; 2006. p. 750–63.
[23] Qooxdoo open source AJAX framework. <http://qooxdoo.org/>.
[24] Zhou Z, Wu Z, Wang C, et al. Efficiently mining both association and correlation rules. Proceedings of the 3rd international conference on fuzzy systems and knowledge discovery, September 24–28, Xi'an, China; 2006. p. 369–72.
[25] Deng S, Wu J, Li Y, et al. Service matchmaking based on semantics and interface dependencies. Proceedings of the 7th international conference on web-age information management, June 17–19, Hong Kong, China; 2006. p. 240–51.
[26] Halevy AY. Answering queries using views: a survey. VLDB J 2001;10:270–94.

[27] Chen H, Wu Z, Wang H, et al. RDF/RDFS-based relational database integration. Proceedings of the 22nd international conference on data engineering, August 20–23, Atlanta, GA; 2006. p. 94.

[28] Wu D, Parsia B, Sirin E, et al. Automating DAML-S Web services composition using SHOP2. Proceedings of the 2nd international Semantic Web conference, October 20–23, Sanibel Island, FL; 2003. p. 195–210.

7 Ontology Development for Unified Traditional Chinese Medical Language System[1]

7.1 Introduction

Traditional Chinese medicine (TCM) is a complete system of medicine encompassing the entire range of human experience. Thousands of scientific studies that support traditional Chinese medical treatments are published yearly in journals around the world. However, even patients who benefit from treatments such as acupuncture or Chinese herbal therapy may not understand all the components of the TCM system. This may be, in part, because TCM is based on a dynamic understanding of energy and flow that has more to do with Western physics than Western medicine. TCM embodies rich dialectical thoughts, such as the holistic connections and the unity of Yin and Yang. The ideas of integration and Bian-Zhen-Lun-Zhi are the fundamental infrastructure of TCM [1].

With the development of information technology and wide use of the Internet, an immense amount of disparate isolated medical databases, electronic patient records (EPR), hospital information systems (HIS), and knowledge sources were developed. In 2000, we developed a unified web-accessible multidatabase query system of TCM bibliographic databases and specific medical databases to address the distributed, heterogeneous information source retrieval in TCM. It has been available on the website for registered users online for 5 years [2]. As a complete system with complex disciplines and concepts, TCM has the main obstacle that large amounts of ambiguous and polysemous terminologies exist during the information processing procedure. We initiated the unified traditional Chinese medical language system (UTCMLS) project in 2001, which is funded by the China Ministry of Science and Technology to study the terminology standardization, knowledge acquisition, and integration in TCM. We recognized that there are three main challenges in the UTCMLS project:

1. To design a reusable and refinable ontology that integrates and accommodates all the complex TCM knowledge.

[1] Reprinted from Zhou X, Wu Z, Yin A, Wu L, Fan W, Zhang R. Ontology development for unified traditional Chinese medical language system. Artif Intell Med 2004;32:15–27. © 2004 Published by Elsevier B.V., with permission from Elsevier.

Modern Computational Approaches To Traditional Chinese Medicine. DOI: http://dx.doi.org/10.1016/B978-0-12-398510-1.00007-8

2. To harness a broad collaboration of different domain experts in distributed sites in ontology development.
3. To develop a knowledge infrastructure for the Semantic Web.

This chapter mainly addresses the former two challenges, which are relevant to the design and development of TCM ontology. Ontology building is a nontrivial and valuable work for domain data, information, and knowledge integration, especially for a complicated and comprehensive domain like TCM. Using Protégé 2000, we try to facilitate the development of TCM ontology and alleviate the labor through methodology and structure design of ontology. The rest of this chapter is arranged as follows. To illuminate TCM as a complete system and to clarify the methodology of ontology design, Section 7.2 gives a detailed overview of the knowledge system of TCM from the discipline perspective. An ontology overview is proposed in Sections 7.3 and 7.4 that introduces Protégé 2000, the ontology tool we use to build the UTCMLS. Section 7.5 discusses the methodology, knowledge acquisition, and information integration for ontology development. Section 7.6 gives the current main results of ontology development. Finally, concluding remarks and future work are proposed.

7.2 The Principle and Knowledge System of TCM

TCM is a medical science that embodies Chinese culture and philosophical principles, which is the basis and essence of the simple and naive materialism in China. Because the understanding of the TCM concept, theory, and philosophy is vital to this ontology's development, we give a brief overview of the TCM knowledge system.

TCM embodies rich dialectical thoughts, such as that of the holistic connections and the unity of yin and yang. It deals with many facets of human anatomy and physiology: zang-fu (organs); meridians (main and collateral channels); qi (vital energy); blood; jing (essence of life); body fluid; the inside and outside of the body; as well as the connections between the whole and the parts. It also examines the effect of the social and natural environment—the universe, the sun and the moon, the weather, the seasons, and geography on the interrelations and conditioning of yin and yang. The result has been the formation of a system of thought about the interrelations behind spirit and organism, zang and fu, and the inside and outside of the body. TCM uses the interrelationship among formations, factors, and variables, both within and outside the body; it regards and deals with these interrelations with reference to data that are correspondingly interrelated; it uses the principle of stabilization to "harmonize yin and yang to reach a state of equilibrium," adjust their relationship so that they remain in a healthy state. The detailed discussions and remarks about the methodology of TCM can be found in Refs. [1,3].

Given the difficulties of the TCM concept (i.e., its complexity, vast, variable, and nonstandard), and the importance given to that enormous amount of ancient literature, which has been a general and core knowledge source in TCM, it is a great challenge and central role to develop an ontology of formally specified concepts and relationships for UTCMLS. The design, implementation, and usage will be described in detail in the next sections.

7.3 What Is an Ontology?

Ontology is a branch of philosophy concerned with the study of what exists. "Ontology" is often used by philosophers as a synonym for "metaphysics" [4]. Philosophical ontology is a descriptive enterprise. It is distinguished from the special sciences not only in its radical generality but also in its primary goal or focus: it seeks no predication or explanation but rather taxonomy. Formal ontologies have been proposed since the eighteenth century, including recent ones such as those by Carnap [5] and Bunge [6].

It was McCarthy [7] who first recognized the overlap between work done in philosophical ontology and the activity of building the logical theories of artificial intelligence (AI) systems. McCarthy affirmed in 1980 that builders of logic-based intelligent systems must first "list everything that exists, building an ontology of our world." According to Gruber [8], an ontology is a "specification of a conceptualization," while Guarino [9] argued that "an ontology is a logical theory accounting for the intended meaning of a formal vocabulary." Ontologies are essential for developing and using knowledge-based systems. Every knowledge model has an ontological commitment [10], i.e., a partial semantic account of the intended conceptualization of a logical theory. Ontologies form the foundation for major projects in knowledge representation such as CYC [11], TOVE [12], KACTUS [13], and SENSUS [14]. Medicine has been the active ontology research and construction area for large knowledge bases. There are several distinguished efforts in medical terminology systems like SNOMED-RT [15] and "Canon group" [16]. The semantic network in unified medical language system (UMLS) [17] is also considered a distinguished terminological ontology [18]. GALEN [19] is a project developing medical terminology servers and data entry systems based on ontology, the common reference model, which is formulated in a specialized description logic, GRAIL [20]. Also, reusable medical ontologies are strongly recommended by Schreiber and Musen [21,22] in intelligent systems.

Van Heijst et al. [23,24] have a case study in the construction of a library of reusable ontologies and they proposed several important and useful principles to address the corresponding hugeness problem and the interaction problem. TCM is a specific domain with a large amount of knowledge. The goal of TCM ontology development is to facilitate the development of TCM terminological knowledge-based system (KBS) by providing a reusable core generic ontology and relevant skeleton subontologies.

7.4 Protégé 2000: The Tool We Use

We use Protégé 2000 [25] as an ontology editor (also as knowledge acquisition tool) with RDFS as the underlying representation language. Protégé 2000 is a

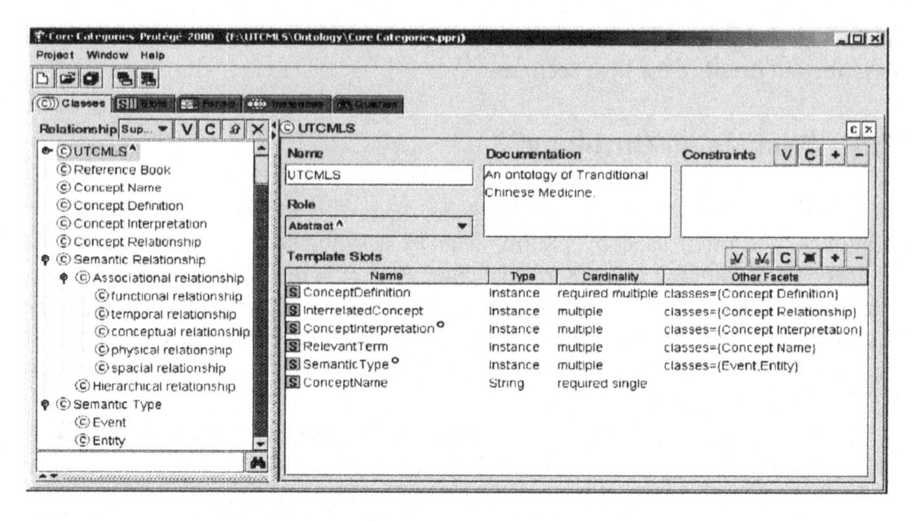

Figure 7.1 Protégé 2000 user interface and the overview of TCM ontology in editing.

frame-based knowledge base development and management system that offers classes, slots, facets, and instances as the building blocks for representing knowledge. The Classes Tab is an ontology editor, which designs classes in flexible style and organizes classes as a hierarchy. Classes have slots whose value may or may not be inherited. Facets specify the cardinality and data type of the slot value. The Instances Tab can help the user acquire the knowledge of a domain. The ontology plus instances can be viewed as a domain knowledge base. We chose Protégé 2000 because:

1. It integrates ontology editors and knowledge acquisition tools in a single application to facilitate the knowledge engineering process;
2. It has the extensible component architecture;
3. It defines a flexible meta-class architecture and supports many formats such as OKBC, RDF/RDFS, and database storage [26], and the DAML + OIL is also supported by the additional plug-ins in the current version.

There is huge amount of knowledge and data that is required to be put into a knowledge base. Protégé 2000 user interfaces are intuitive for domain experts to work with for knowledge acquisition. Figure 7.1 is an interface screen shot for ontology development in Protégé 2000. As can be seen on the left frame of Protégé interface, we defined six core top-level classes, namely semantic type, semantic relationship, concept name, concept definition, concept interpretation, and concept relationship in TCM ontology. The right frame contains the specific definitions of a class (i.e., name, documentation, constraints, role, and template slots). Section 7.5 gives the detailed description of the six core top-level classes definitions.

7.5 Ontology Design and Development for UTCMLS

The development of ontologies is a modeling activity that needs the ontological engineers (also called ontologists) who have sufficient understanding of the domain of concern and are familiar with knowledge representation languages. Problems like ontology and conceptual modeling need to be studied in a highly interdisciplinary perspective [27]. Ontology construction is a complex and labor-intensive activity. Several controlled vocabularies and many special terminology lexicons exist in TCM, but combining and synchronizing individual versions of existing medical terminology vocabularies into a unified terminological system is still a problem, because of the heterogeneity and indistinctness in the terminology used to describe the terminological systems [28]. It is accelerated by the various and nonstandard use of words in the clinical practice. The ontology development for UTCMLS is still in the preliminary stage. Only a small part of subontologies (e.g., the basic theory of TCM, formula of herbal medicine, Chinese materia medica, and acupuncture) has been developed. About 8000 class concepts and 50,000 instance concepts are defined in the current TCM ontology, whereas we estimate the number of concepts of TCM will be up to several hundreds of thousands, maybe even reach several millions. Furthermore, because the terminology knowledge is mastered and used in practice by different groups of experts, there should be a broad cooperation in the knowledge acquisition procedure. A methodology of loosely coupled development and quality assurance is needed to build a formal final ontology. Although agreement on a high-level schema is a prerequisite for effective cooperation between different groups of modelers, it is not practically possible to build a philosophically perfect model. We would rather aim to ensure the ontology fits closely enough with most usage requirements to be refinable and to be reusable. We give a summary discussion of the methodology, knowledge acquisition, design, and development of ontology in TCM in the successive sections.

7.5.1 Methodology of Ontology Development

The use and importance of ontologies is widespread. However, building ontologies is largely a black art. All the methodologies, such as TOVE [12] and SENSUS [15], are task specific. Uschold [29] emphasized that no unified methodology is suitable for all the jobs, but different approaches are required for different circumstances. Jones et al. [30] had an overview of ontology methodologies and proposed guidelines for ontology development. It shows that ontological engineering is still a craft rather than a science at present. There are two main tasks involved in content development for the TCM ontology: (1) knowledge acquisition and conceptualization from disparate TCM information sources and (2) formulization and implementation of ontology schema. The building of TCM domain ontology conforms to several principles.

1. Refinement is needed. The methodology, which is based on the prototype ontology, is preferred in TCM ontology building for the vast, complex, and dynamic knowledge system in TCM.
2. Development should be based on informal ontologies. There are many terminology lexicons and a number of controlled vocabularies. These knowledge sources can be viewed as the informal ontologies at the start point.

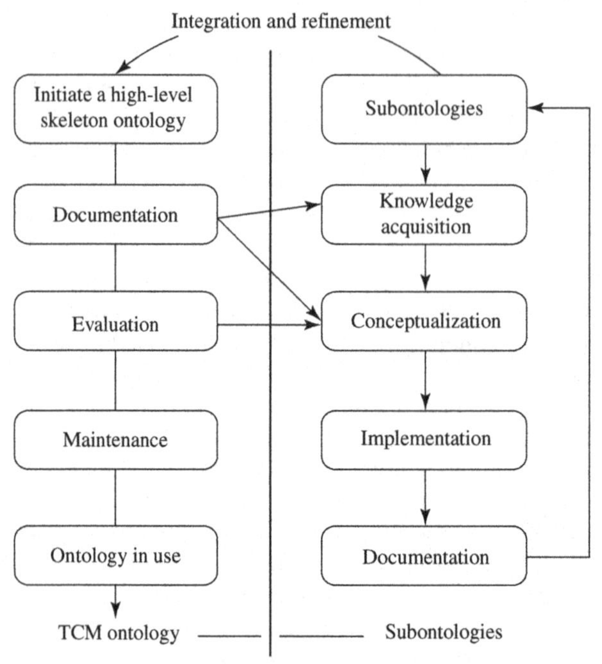

Figure 7.2 The cooperative loosely coupled development of TCM ontology.

3. Evaluation is essential and important. Ambiguity and polysemia are the characteristic phenomena of TCM concepts. No standard has been agreed on the concept structure and relationship in the TCM discipline. The ontology development of TCM is a continuous procedure of evaluation and development.
4. The methodology of distributed loosely coupled development is required. The task of building TCM ontology is distributed across 16 sites in China, and each site is responsible for corresponding subdomain terminology work of TCM.

Rector et al. [30] had the practice of distributed cooperative ontology development in the medical domain, in which an intermediate representation at the knowledge level has been used to control the quality of ontology development. We also adopt the intermediate representation mechanism based on tabular and graph notations. The knowledge acquisition, conceptualization, integration, implementation, evaluation, and documentation activities are involved in the ontology development. The evaluation mainly focuses on the quality control of the intermediate representation from the distributed domain experts. We also make guidelines and the criteria of Gruber [8] for domain experts to control the process of knowledge acquisition and conceptualization. According to the above principles, we applied development criterion to both whole TCM ontology and individual subontologies, as shown in Figure 7.2. The current practice showed that this criterion assured the quality control of ontology, interaction, and communication between domain experts and a central knowledge engineering team.

A core generic framework of TCM ontology is designed at the start point of ontology development. About 14 subontologies and six core top-level classes

(Figure 7.4 depicts the definitions of each class) are defined as the initial skeleton ontology. The subontologies are defined according to the disciplines of TCM based on most domain experts' viewpoints, as shown in Table 7.1.

7.5.2 Knowledge Acquisition

In 2001, the China Academy of Traditional Chinese Medicine and College of Computer Science, Zhejiang University, initiated and organized the project of UTCMLS as the basis of all the TCM information-related work. We aim to implement a unified TCM language system, which stimulates the knowledge and concept integration in TCM information processing. The National Library of Medicine in the United States has assembled a large multidisciplinary, multisite team to work on the UMLS [17], aimed at reducing fundamental barriers to the application of computers to medicine [31].

UMLS is a successful task in medical terminology research, which inspired our research of TCM knowledge and terminology problems. Much good experience has been learned from UMLS. The structure of TCM ontology is heavily influenced by the semantic network of UMLS. However, the work of TCM as a complete discipline system and complex giant system is much more complicated. Some principles should be adhered to during the knowledge acquisition procedure.

1. *Deep analysis of specialized lexicons as the knowledge source*: The scientific control of the conceptual glossary of a discipline is the most important issue in natural and artificial language processing. We combine the precontrolled vocabulary and postcontrolled vocabulary as a whole and have a multilevel description of a conceptual glossary such as containing morphology, lexics, semantics, and pragmatics.
2. *A good construction of the TCM-oriented concept framework*: TCM involves the complete discipline system including medicine, nature, agriculture, and humanities. The knowledge system of TCM has complex semantic concept structures, types, and relationships, which refer to multidiscipline content. Therefore, comprehensive analysis and research of the terminology of TCM is needed before the ontology design. A TCM-oriented concept framework should be constructed to address all the knowledge engineering problems of related disciplines.
3. *Efficient combination of controlled vocabularies and specialized lexicons*: As TCM has various concept taxonomical frameworks, a unified concept infrastructure should be established on the basis of controlled vocabularies and specialized lexicons. The controlled vocabulary can be viewed as an ontology, which has no instance. The specialist lexicon can be viewed as the instance of ontology. Both of them constitute a knowledge base.
4. *Development of TCM science and other relevant disciplines*: We develop the UTCMLS not only for information integration and processing in the TCM field but also for agriculture, pharmaceutical technology, and Western scientific medicine. This is coordinated along with the characteristics of the TCM discipline system.

From the perspective of the TCM discipline, considering the medical concept and its relationship, we define the TCM knowledge system by two components: concept system and semantic system. The concept system initially contains about

Table 7.1 The Initial Indispensable Subontologies Defined Corresponding to the Disciplines of TCM

Subontologies	**Characterization of Content**
The basic theory of traditional Chinese medicine	Defines TCM basic theoretical notions such as yin yang, five elements, symptoms, etiology, pathogenesis, physiology, and psychology
The doctrines of traditional Chinese medicine and relevant science	Defines basic notions about the doctrines of TCM, which is based on Chinese ancient philosophy and TCM clinical practice. The concept knowledge of TCM-relevant science is also defined
Chinese materia medica (herbal medicine)	Contains the natural medicinal materials used in TCM practice such as plants, animals, and minerals
Chemistry of Chinese herbal medicine	Contains basic notions of chemical ingredients such as rhein, Arteannuin, and Ginsenoside RB2, which are distilled, separated, identified from herbal medicine, and their structures are also measured
Formula of herbal medicine	Defines basic notions such as benzoin tincture compound, rhubarb compound, and Cinnabar compound, which are based on the theory of prescription of formula
Acupuncture	Defines basic notions of the modern acupuncture discipline, which is on the basis of thoughts of ancient acupuncture and uses traditional and modern techniques to study the issue of meridians, point, rules of therapy, and mechanism
Phamaceutics and agriculture	Defines basic concepts of medical techniques in the manufacture and process of medicinal materials and ontological categories of agriculture issues relevant to medicine planting
Humanities	Defines basic notions about terminology interpretation and relevant knowledge of TCM ancient culture and TCM theories
Informatics and philology	Contains basic notions of TCM-relevant informatics and philology
Medicinal plants and other resources	Defines the medicinal plants and other resources, which are used in health care and disease prevention/cure
Other natural sciences	Defines basic notions of TCM-relevant disciplines, which study natural and physical phenomena
Prevention	Defines basic notions of the science to prevent the occurrence and development of disease
Administration	Defines basic notions of medical research organizations and relevant administration
Geography	Contains the basic notions (e.g., toponym, climate, and soil) that are relevant to TCM

Each subclass of subontologies is mainly defined by the different experts of institutions. We have also established a nomenclature committee to evaluate the definitions.

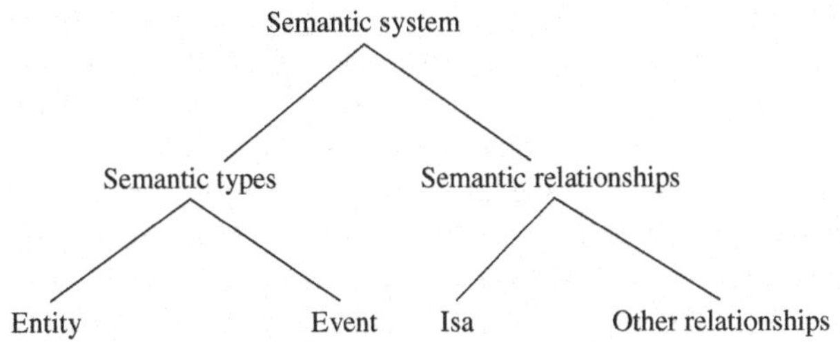

Figure 7.3 The semantic system framework. The detailed class definitions and instances of semantic types and semantic relationships are provided in Section 7.6.

14 subontologies according to the division of the TCM discipline and four basic top-level classes to define each concept. The semantic system concerns the semantic type and semantic relationship of the concept (Figure 7.3).

According to the time, function, space, entity, and concept attributes of TCM knowledge, we have defined 59 kinds of semantic relationships between concepts and about 104 kinds of TCM semantic types, plus all the semantic types of UMLS. In Section 7.6, we will introduce the current version of TCM ontology.

7.5.3 Integrating and Merging of TCM Ontology

Information integration is a major application area for ontologies. Ontology integration is possible only if the intended models of the original conceptualizations of the two ontologies are associated with overlap [9]. In the procedure of TCM ontology development, we must let the ontology be built by different experts in a distributed environment for the very complex knowledge acquisition work of TCM. We use the top-down approach to developing the 14 subontologies and the other six core top-level classes and distribute the 14 subontologies to the domain experts of about 16 TCM research organizations in China. The bottom-up approach is used during the development of each subontology. Therefore, ontology merging (information integration) is a must. We use IMPORT to merge the subontologies from different sources to a unified TCM ontology. IMPORT is a plug-in of the Protégé 2000, which is the latest version of SMART [32]. It is shown from TCM ontology practice that IMPORT is an effective tool to merge ontologies.

7.6 Results

The development of UTCMLS is a process of building a systematized general knowledge-oriented TCM terminological system through an ontology approach. We have a nomenclature committee consisting of seven TCM linguistic and terminological experts to evaluate the nomenclature standard and fix the final definitions with the other participant linguistic and domain experts. More than 30 experts in the fields of TCM, medical informatics, knowledge engineering, and medical administration were consulted about the development of TCM ontology. The categories of the structure of the TCM ontology are formed according to the structures of the controlled vocabularies and the standard textbooks. More than 100 controlled vocabularies and terminologies have been chosen as the sources for TCM ontology, which are stored in a simple ontology-named reference book ((8) in Figure 7.4). Some of the reference books used as main knowledge sources are *Chinese Traditional Medicine and Materia Medical Subject Headings* [33], *Traditional Chinese Medical Subject Headings* [34], *Chinese Library Classification* (4th ed.) [35], *National Standard* [36–38], and *Pharmacopoeia of the People's Republic of China* [39].

Based on these existing controlled vocabularies and terminologies, manual knowledge distilling and knowledge extraction are the approaches taken to handle concept definition and organization. A basic principle we conform to is that the three controlled vocabularies, namely, Chinese Traditional Medicine and Materia Medica Subject Headings, Traditional Chinese Medical Subject Headings, and Chinese Library Classification (4th ed.), are considered as the main knowledge sources of ontology, but we prefer to use National Standard and Clinical Diagnosis and Treatment Terms (e.g., Classification and codes of diseases and ZHENG of traditional Chinese medicine/GB/T 15657—1995; Clinic terminology of traditional Chinese medical diagnosis and treatment—diseases, syndromes, therapeutic methods/GB/T 16751.1—1997, GB/T 16751.2—1997, GB/T 16751.3—1997) when the terminology definitions of the above three controlled vocabularies are in conflict with those of the two terminological systems. Then the final definitions are defined by the participant domain experts and nomenclature committee. The translation is done from those sources to the TCM ontology by building the relations between terms. Sixteen medical information institutes or medical libraries joined the research group to establish the principles and rules for TCM ontology development as well as to build the ontology. The translation from the sources to the TCM ontology was done according to the following principles. The relationships between terms were built based on the concepts. Different terms from various sources with the same concepts were connected in this way. The synonyms in different forms were translated by the system into the corresponding subject headings. All the terms with the same concepts were selected from the sources first, and then the relationships between terms were built. The nomenclature committee and experts defined the subclasses of each category in TCM ontology. There were some intensive debates. For example, there were 12 subclasses in the category of the basic theory of TCM in the first draft of the structure, but some experts did not agree with this classification. After the discussion, a new structure with six subclasses was developed. The UTCMLS project is still in progress. The core top-level categories, such as the concept-relevant categories, semantic type, and semantic relationship, and the first-level subclass definitions of 14 essential subontologies are currently finished. Furthermore, the complete ontology definitions and knowledge acquisition of some subontologies (e.g., the basic theory of TCM, acupuncture, and formula of herbal medicine) have also been completed. This section provides the current main results and experience of ontology development by introducing the whole framework of the skeleton top-level categories of TCM ontology and the

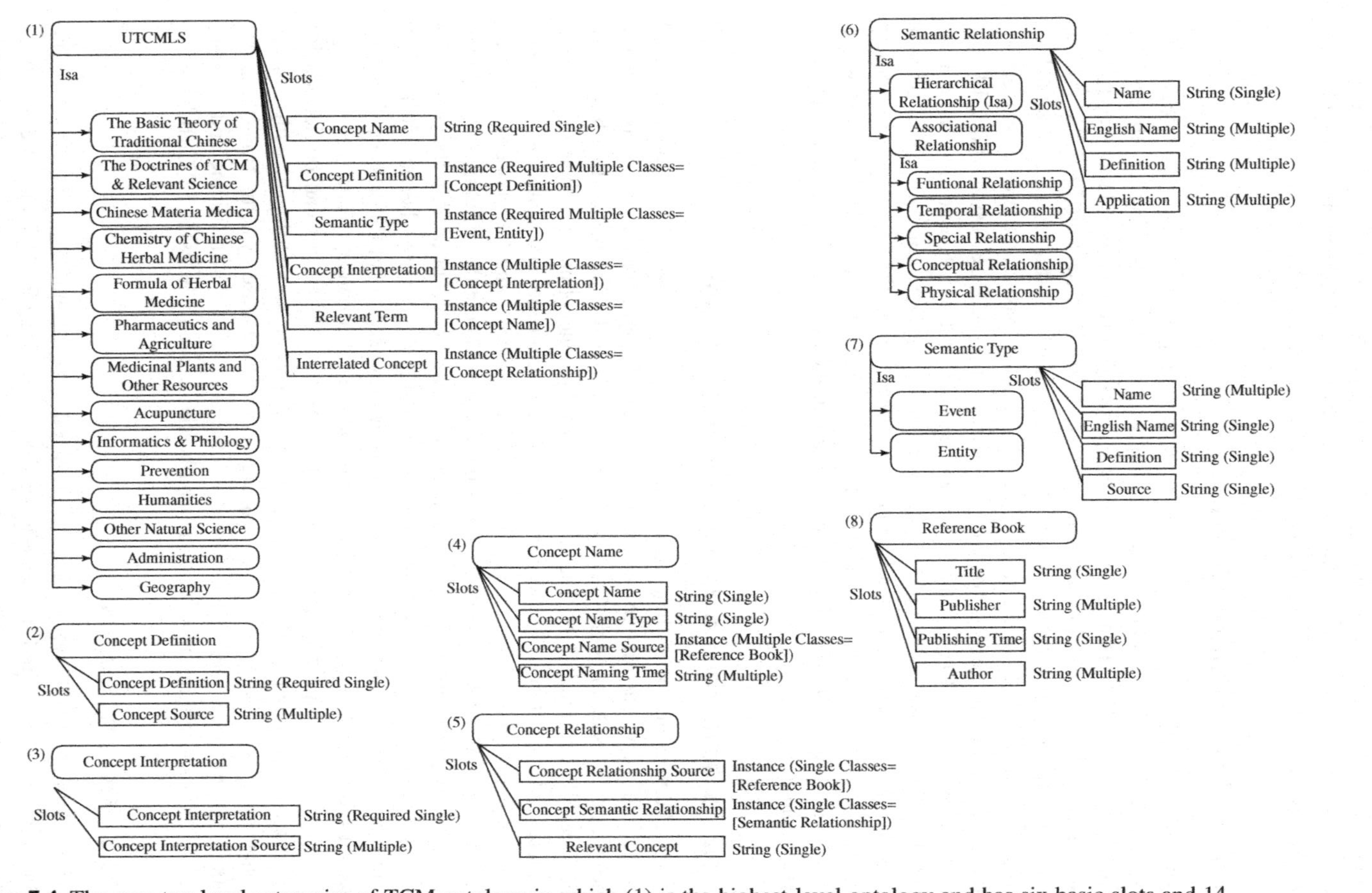

Figure 7.4 The core top-level categories of TCM ontology in which (1) is the highest-level ontology and has six basic slots and 14 subontologies; (2), (3), (4), (5) constitute the concept system; (6), (7) form the semantic system; and (8) is a simple ontology of reference book.

semantic types. Figure 7.4 shows the whole framework of the skeleton top-level class definitions of TCM ontology in Protégé 2000 (RDFS is the underlying knowledge representation language and storage format).

7.6.1 The Core Top-Level Categories

To unify and initiate the whole ontology development and knowledge acquisition, we have defined the core top-level categories of TCM ontology with essential instances (e.g., 104 TCM semantic types and 59 semantic relationships). The semantic types and relationships are depicted in detail in Section 7.6.2.We provide six core top-level categories, which are treated as the instances of the metaclass: STANDARD-CLASS in Protégé 2000, and on the basis of them define all the intentional and extensional content of concepts in UTCMLS. Meanwhile, 14 subontologies with first-level subclasses are defined, but the second or deeper level subclass definitions are mainly determined by the corresponding groups of domain experts who take charge of the knowledge acquisition work, because the bottom-up method is used to facilitate the knowledge acquisition and to decrease the workload. As shown in Figure 7.4, three basic components constitute the core top-level TCM ontology.

7.6.2 Subontologies and the Hierarchical Structure

The subontologies and their hierarchical structure reflect the organization and content of TCM knowledge. The initial 14 subontologies (Table 7.1 illustrates the definitions of subontologies and their contents) and the six basic slots have been defined ((1) of Figure 7.4). The six basic slots are concept name, concept definition, concept interpretation, relevant term, interrelated concept, and semantic type. The concept name slot gives the standard name of a concept, and the concept definition slot and interpretation slot give the descriptive content about the meaning of a concept. The relevant term slot defines the different terms of a concept used in the other relevant vocabulary sources, by which we can construct the relations between UTCMLS and other terminological sources. The interrelated concept slot defines the concepts, which has some kinds of semantic relationships with a concept. The semantic type slot gives the semantic type definition of a concept. More slots can be defined for an individual subontology if necessary. The complete subontologies with concept instances will become a medical language knowledge base system. Now there are 8000 concepts (e.g., herbal medicine and chemistry of herbal medicine and disease) and 50,000 concept instances (e.g., Rhubarb, Rhein, and diabetes) in UTCMLS.

7.6.3 Concept Structure

Using Protégé 2000, we provide a knowledge representation method to define the concept in a unified mode. We consider that every TCM concept consists of three basic intentional attributes, namely, definition, interpretation, and name, hence three classes, namely, concept definition, concept name, and concept interpretation ((2), (3), (4) in Figure 7.4), are defined to construct a terminological concept. The class of concept definition involves the definitions of essential meanings of a concept. The class of concept interpretation gives the explanation of a concept. The class of concept name defines the synonyms, abbreviations, and lexical variants of a concept, that is to say the concept name gives the relevant terminological names of a concept from different controlled vocabularies. Together with the concept name slot of each subontology, these three classes give the lexicon level knowledge of a concept. However, the semantic structure aims at the semantic-level knowledge of a concept.

7.6.4 Semantic Structure

The semantic type and relationship classes ((6), (7) in Figure 7.4) form the foundation of semantic-level knowledge of a concept. The semantic types provide a consistent categorization of all concepts represented in UTCMLS, and the semantic relationships defining the relations may hold between the semantic types. Classification and inference are supported by these definitions of semantic type and relationship of a concept. As Figure 7.4 shows, we define semantic relationship as a slot of concept relationship class and semantic type as an essential slot of TCM ontology to assign the semantic content to concept. Each concept of UTCMLS is assigned at least one semantic type. In all cases, the most specific semantic type available in the hierarchy is assigned to the concept. The semantic structure enables us to construct an abstract semantic network of all the concepts in UTCMLS, so it is vital to TCM ontology. This chapter will list all the TCM-oriented semantic types in Section 7.6.2.

7.6.5 Semantic Types and Semantic Relationships

The semantic network of UMLS [17] is a high-level representation of the biomedical domain based on semantic types under which all the Metathesaurus concepts are categorized. The semantic network makes UMLS a unified medical language system, which is different from other classification and terminological systems. We define the semantic structure in UTCMLS to construct the semantic-level knowledge framework of the TCM concept from the idea of the semantic network of UMLS. We define the semantic type and semantic relationship as one of the two top-level categories of TCM ontology. The structures of semantic type and semantic relationship are defined as shown in Figure 7.4. Most of semantic relationships in TCM ontology are same as UMLS, but there are five more semantic relationships in TCM ontology than in UMLS, namely, being exterior–interiorly related, produces, transforms to, restricts, and cooperates with. Those are special ones used in the unique system of TCM. For example, being exterior–interiorly related is the relationship between the concepts with zang-fu semantic type, which is a special semantic category in TCM. We have defined 104 semantic types (including 40 entity definitions and 64 event definitions) to describe the high-level concept categories in TCM. The 40 TCM-oriented entity semantic types and the hierarchical structure are depicted in Figure 7.5, and the 64 TCM event semantic types are

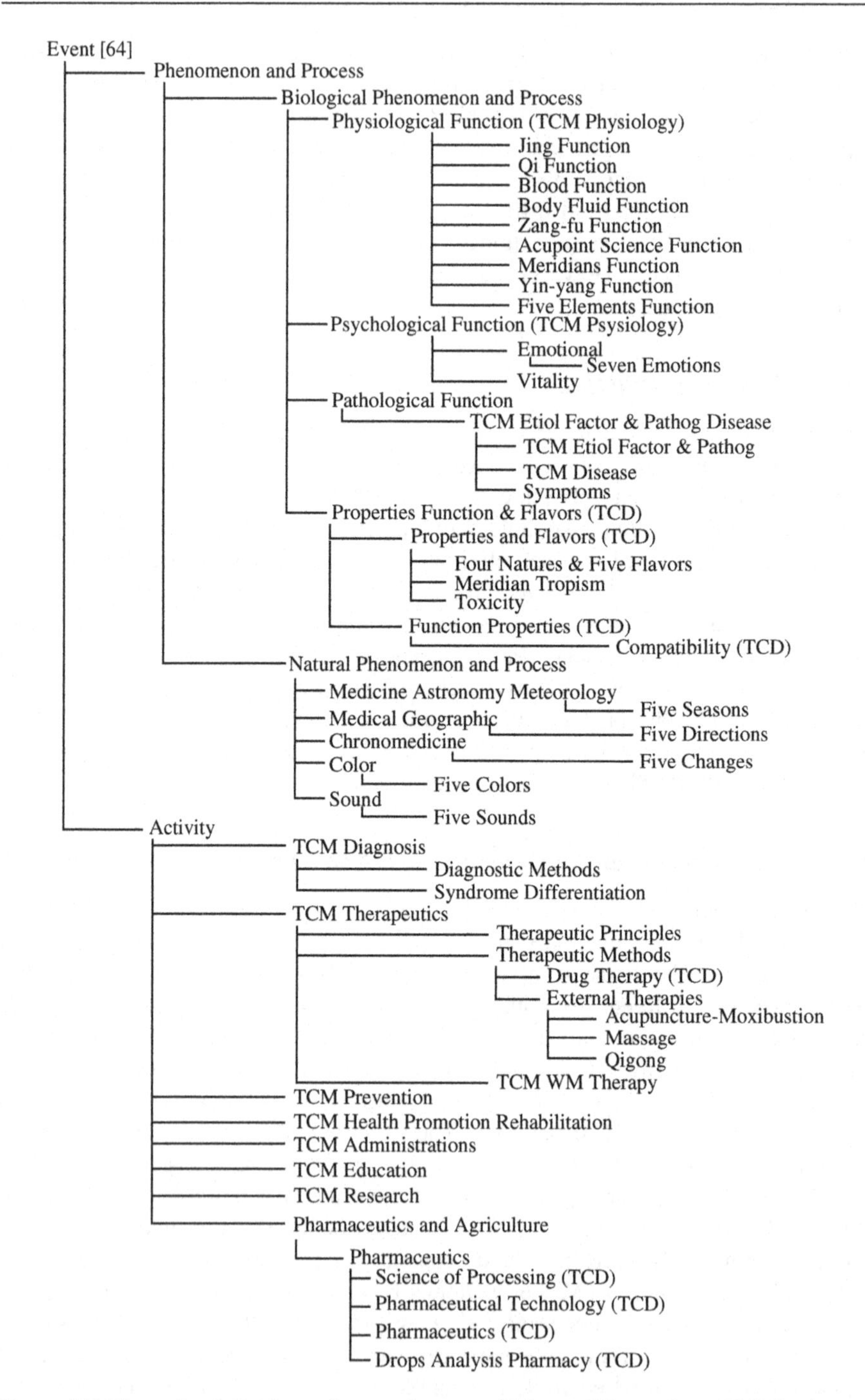

Figure 7.5 The entity definitions of semantic type, which are different from UMLS.

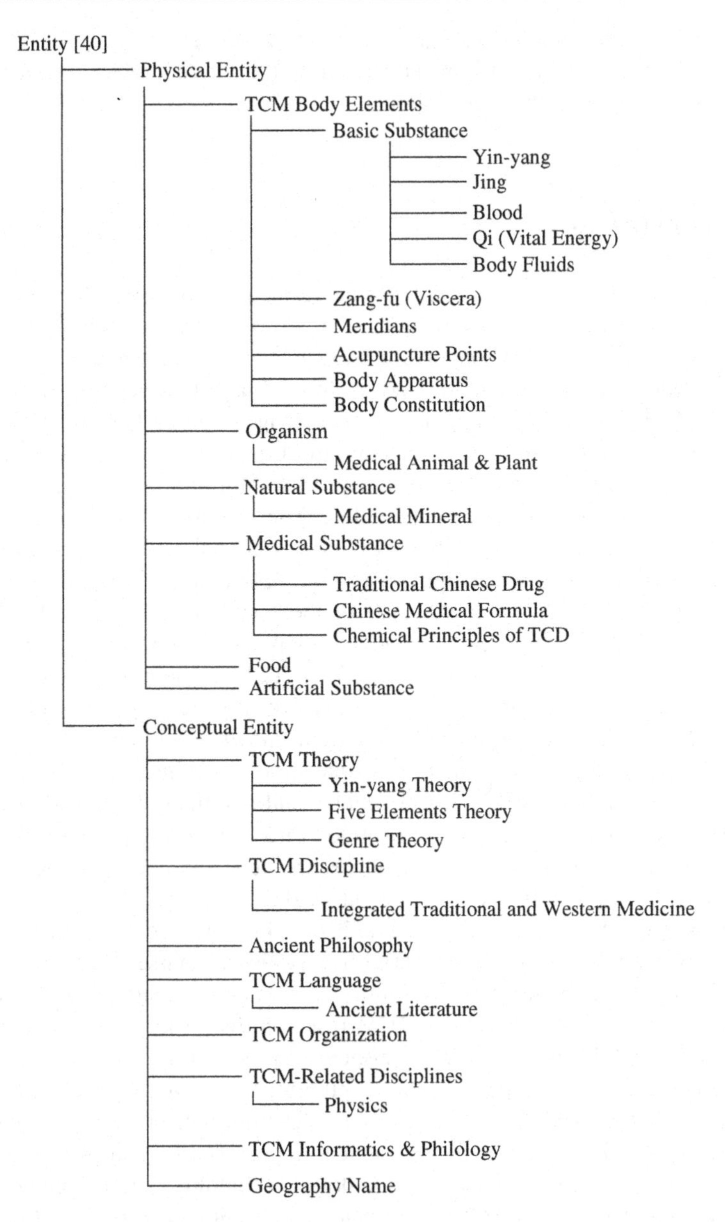

Figure 7.6 The 64 TCM special event definitions of semantic type. The definitions such as Yin-yang function, Zang-fu function, and seven emotions fully reflect the characteristics of TCM terminological knowledge.

listed in Figure 7.6. All the definitions of the semantic system were finally fixed by intensive debates and rigorous evaluation. However, incompleteness and imperfection is allowed for in practice. The detailed definitions of semantic types and relationships and the rules of their applications to concepts are given in the technical report [40] and are also contained in the core top-level ontology of TCM. The whole ontology will be published and shared on the Internet when finished, so this article does not give further descriptions.

7.7 Conclusions

In the current literature on knowledge management, it is often observed that the main challenges are in the realm of human organizational culture and practices. The key to providing useful support for knowledge management lies in how meaning is embedded in information models as defined in ontologies. In China, various kinds of TCM terminology systems with differing purposes have been developed during the past two decades. Unfortunately, most of them are paper based that cannot satisfy anymore the desiderata of health care information systems, such as the demand for reuse and sharing of patient data. The unambiguous communication of complex and detailed medical concepts is now a crucial feature of medical information systems. An ontological approach to the description of terminology systems will allow a better integration and reuse of these systems. Ontology development for the UTCMLS is a systematic and comprehensive procedure of knowledge acquisition and knowledge integration in TCM. TCM embodies knowledge of systematics, cybernetics, and informatics and involves a wide range of multidiscipline terminologies. It is tempting to build a unified knowledge-oriented terminological system with consistent, formal, and extensible structure to integrate the existing terminology systems, but it will be a huge labor- and intelligence-intensive work, like the UMLS project. Sixteen TCM institutes and colleges with several hundred researchers have been involved in this project to build the ontology and to acquire the knowledge. To satisfy the requirement of distributed development, a scalable and extensible methodology is focused on in TCM ontology development. The process of ontology development for UTCMLS is also a course of medical concept standardization and unification. Due to the vast knowledge storage and very complex knowledge system involved in TCM research, the process of ontology development for UTCMLS is a continuous refinement procedure.

More precisely, this chapter presents a preliminary ontology development experience of the TCM medical language system. The main effort of this chapter is to provide an ontology approach to standardizing TCM medical terminology. Furthermore, the TCM-oriented methodology of ontology development and an ontology structure are defined to facilitate the development of UTCMLS KBS. According to the knowledge system of TCM, two subsystems (e.g., concept system and semantic system) are defined to describe and construct the terminological system. The structure of the semantic system is inherited from the semantic network of UMLS, but many more semantic types (about 104 additional semantic types) are defined to reflect the essentials of TCM science. Four core top-level categories, namely, concept definition, concept name, concept interpretation, and concept relationship, constitute the concept system, which build a unified approach to defining the TCM concept. We aim to build a unified ontology framework to effectively organize and integrate the terminological knowledge of TCM. Although the core top-level categories and essential ontology structures are defined, much more work (e.g., terminology standardization, problem-solving methods, and knowledge acquisition) should be done to construct a final version of TCM ontology and medical terminology system, which will support various medical applications.

Now the applications of UTCMLS, such as concept-based medical information retrieval and TCM-specific semantic browser (exploring the information and structure of UTCMLS online), are in progress. Future work on UTCMLS includes: (1) completion of TCM ontology and expansion of the subontologies; (2) completion of knowledge acquisition to build a final medical language system; (3) interrelating to the existing medical database sources; (4) development of a TCM Semantic Web terminology server; (5) development of health care information systems based on the UTCMLS terminology server.

References

[1] Yin HH, Zhang BN. The basic theory of traditional Chinese medicine. Shanghai: Shanghai Science and Technology; 1984 [in Chinese].

[2] <http://www.cintcm.com/>.

[3] Huang JP. Methodology of traditional Chinese medicine. Beijing: New World Press; 1995.

[4] Mealy GH. Another look at data. Proceedings of the AFIPS fall joint computers conference. Washington, DC: Thompson Book Co.; 1967. p. 525–534

[5] Carnap R. The logical structure of the world: pseudo-problems in philosophy. California: University of California Press; 1967.

[6] Bunge M. Treatise on basic philosophy: ontology I: the furniture of the world. New York, NY: Reidel; 1977.

[7] Smith B, Welty C. Ontology: towards a new synthesis. In: Welty C, Smith B, editors. Formal ontology in information systems. New York, NY: ACM Press; 2001. p. 3–9.

[8] Gruber TR. Toward principles for the design of ontologies used for knowledge sharing. Int J Hum Comput Stud 1995;43:907–28.

[9] Guarino N. Formal ontology and information systems. In: Guarino N, editor. Formal ontology in information systems. Amsterdam: IOS Press; 1998. p. 3–15.

[10] Noy N, Hafner C. The state of the art in ontology design: a survey and comparative review. AI Mag 1997;18:53–74.

[11] Lenat DB. CYC: a large-scale investment in knowledge infrastructure. Commun ACM 1995;38:33–38.

[12] Uschold M, Gruninger M. Ontologies: principles, methods and applications. Knowledge Eng Rev 1996;11:93–155.

[13] Schreiber G, Wielinga B, Jansweijer W. The KACTUS view on the "O" word. Proceedings of the workshop on basic ontological issues in knowledge sharing/international joint conference on artificial intelligence. Menlo Park, CA: AAAI Press; 1995. p. 92–100

[14] Swartout B, Patil R, Knight K, et al. Toward distributed use of large-scale ontologies. In: Proceedings of the AAAI-97 spring symposium on ontological engineering, Stanford, CA; 1997. p. 111–18.
[15] Spackman KA, Campbell KE, Cote RA. SNOMED RT: a reference terminology for health care. In: Proceedings of the AMIA fall symposium; 1997. p. 125–34.
[16] Evans DA, Cimino J, Hersh WR, et al. Position statement: towards a medical concept representation language. J Am Med Inform Assoc 1994;1(3):207–14.
[17] Lindberg DAB, Humphreys BL, McCray AT. The unified medical language system. Methods Inform Med 1993;32:281–91.
[18] Uschold M. Building ontologies: towards a unified methodology. In: Proceedings of the 16th annual conference of the British computer society specialist group on expert systems, Cambridge, UK; 1996.
[19] Rector AL, Nowlan WA, and the GALEN Consortium. The GALEN project. Comput Methods Programs Biomed 1993;45:75–8.
[20] Rector AL, Bechhofer S, Goble C, et al. The GRAIL concept modelling language for medical terminology. Artif Intell Med 1997;9:139–71.
[21] Musen M. Modern architectures for intelligent systems: reusable ontologies and problem-solving methods. Am Med Inform Assoc Symp Suppl 1998;:46–54.
[22] Musen M, Schreiber G. Architectures for intelligent systems based on reusable components. Artif Intell Med 1995;7:189–99.
[23] Van Heijst G, Falasconi S, Abu-Hanna A, et al. A case study in ontology library construction. Artif Intell Med 1995;7:227–55.
[24] Van Heijst G, Schreiber AT, Wielinga BJ. Using explicit ontologies for KBS development. Int J Hum Comput Stud 1997;42:183–292.
[25] <http://protege.stanford.edu/>.
[26] Eriksson H, Fergerson R, Shahar Y, editors. Automatic generation of ontology. In: Proceedings of the 12th Banff knowledge acquisition workshop, Banff, Alberta, Canada; 1999. p. 93–100.
[27] Guarino N. Formal ontology, conceptual analysis and knowledge representation. Int J Hum Comput Stud 1995;43:625–40.
[28] De Keizer NF, Abu-Hanna A, Zwetsloot-Schonk JHM. Understanding terminological systems I: terminology and typology. Methods Inform Med 2000;39:16–21.
[29] Jones D, Bench-Capon T, Visser P. Methodologies for ontology development. In: Proceedings of the IT&KNOWS conference, XV IFIP world computer congress, Budapest; 1998. p. 99–105.
[30] Rector AL, Wroe C, et al. Untangling taxonomies and relationships: personal and practical problems in loosely coupled development of large ontologies. In: Gil Y, Musen M, Shavlik J, editors. K-CAP'01; 2001. p. 139–46.
[31] Humphreys BL, Lindberg DA. The unified medical language system: an informatics research collaboration. Am Med Inform Assoc 1998;5:1–11.
[32] Fridman Noy N, Musen MA. SMART: automated support for ontology merging and alignment. In: Proceedings of the 12th workshop on knowledge acquisition, modelling and management (KAW'99), Banff, Canada; 1999. p. 108–18.
[33] Wu LC (chief editor). Chinese traditional medicine and materia medical subject headings. Beijing: Chinese Medical Ancient Books Publishing; 1996.
[34] Beijing College of Traditional Chinese Medicine. Thesaurus of traditional Chinese medicine. Traditional Chinese medical subject headings. Beijing: Beijing Science and Technology Press; 1987.

[35] Editing Committee of Chinese Library Classification. Chinese library classification. 4th ed. Beijing: Beijing Library Press; 1999.

[36] General administration of technology supervision of the People's Republic of China. National standard: clinic terminology of traditional Chinese medical diagnosis and treatment—diseases. Beijing; 1997.

[37] General administration of technology supervision of the People's Republic of China. National standard: clinic terminology of traditional Chinese medical diagnosis and treatment—syndromes. Beijing; 1997.

[38] General administration of technology supervision of the People's Republic of China. National standard: clinic terminology of traditional Chinese medical diagnosis and treatment—therapeutic methods. Beijing; 1997.

[39] The Pharmacopoeia Commission of PRC. Pharmacopoeia of the People's Republic of China. Beijing: Chemical Industry Press; 1997.

[40] Yin AN, Zhang RE. The blue print of unified traditional Chinese medical language system. Technical report; 2001 [in Chinese].

8 Causal Knowledge Modeling for Traditional Chinese Medicine Using OWL 2

8.1 Introduction

The difference between Western medicine (WM) and traditional Chinese medicine (TCM) is that WM focuses on the anatomy of the human body, whereas TCM is based on an entirely different system of inherent rules. Because of the complicated philosophy, TCM has not been properly understood using current computer technologies.

The primary goal of the Semantic Web [1] is to use URIs as a universal space to name anything, expanding from using URIs for web pages to URIs for "real objects and imaginary concepts," as phrased by Berners-Lee. Among those efforts done by the W3C group, Resource Description Framework (RDF) [2] provides data model specifications and XML-based serialization syntax, Web Ontology Language (OWL) [3] enables the definition of domain ontologies and sharing of domain vocabularies. The OWL 2 Web Ontology Language, informally OWL 2, provides a possible solution for rule reasoning using property chains.

For what concerns us, we wish to apply the Semantic Technologies to represent TCM knowledge, which is mainly focused on rule reasoning derived from TCM philosophy. For example, the Five Phase Theory [4] of TCM divides all the things in the whole world into five types, which are Wood, Fire, Earth, Metal, and Water. The system of five phases was used for describing interactions and relationships between phenomena.

A lot of work has also been done to study the causal relationship in TCM. A stepwise causal adjacent relationship discovery algorithm [5] has been developed to study the correlation between composition and bioactivity of herbal medicine and identify active components from the complex mixture of TCM. The Chinese Medical Diagnostic System (CMDS) [6] contains an integrated medical ontology, and its prototype can diagnose about 50 types of diseases by using over 500 rules and 600 images for various diseases. Wang et al. [7] developed a self-learning expert system for diagnosis in TCM using a hybrid Bayesian network learning algorithm, Naïve-Bayes classifiers with a novel score-based strategy for feature selection and a method for mining constrained association rules. However, these researches focused on applying mathematical methods to TCM knowledge mining and learning, so domain

Modern Computational Approaches To Traditional Chinese Medicine. DOI: http://dx.doi.org/10.1016/B978-0-12-398510-1.00008-X

ontology only acts as a knowledge base without self-learning capability, although building inherent knowledge links can be applicable using current ontology language.

TCM is a static theory model rather than an ever-evolving statistical one, so an expressive causal knowledge model with built-in rules can reveal the nature of TCM better. In this chapter, we present an approach to building a TCM knowledge model with the capability of rule reasoning based on property chains using OWL 2.

8.2 Causal TCM Knowledge Modeling

We implement the causal TCM knowledge model with five functional layers. The ontology layer gives the basic terminology and assertions represented using OWL 2. Association rules are gathered in rule layers. A rule engine layer is recognized as an engine to deal with rules. A method layer does knowledge mining based on the defined ontology and rules. The utility of the causal TCM knowledge model ought to be the natural representation of TCM knowledge; it is also a reference model for TCM knowledge reasoning and knowledge mining, more importantly.

Our TCM knowledge model is designed based on the basic principle of the Five Phase Theory and the aim is to enable causal reasoning of TCM. According to the theory, all the things in the universe can be mapped to one of the five elements known as Wood, Fire, Earth, Metal, and Water. We define seven top ontology classes and corresponding subclasses to enable later causal reasoning.

In the top class design, "Five Phases" refers to the basic five elements in Five Phase Theory. "Environment" defines all kinds of natural elements that we use to diagnose diseases. "Body Elements" include all the physiological components. "Physiology" describes the related descriptions which describe human body conditions. "Pathology" describes pathologic states of the human body. "Treatment" includes Chinese medicine, recipes, rules of treatment, and therapy. "Symptoms" contains abnormal human body states.

In OWL 2, applications need to model interactions that are referred to as one property "propagating" or being "transitive across" another. For now, we define about 30 object properties, which will help us define the property chains in the reasoning stage. For instance, we have: *ObjectPropertyAssertion(:creates:Wood:Fire)* to state *Wood* generates *Fire* in the theory. These internal properties between them give a basic causal foundation of the TCM knowledge model.

As stated above, ontology classes establish a terminology structure for TCM domain knowledge; refined properties connect the terminology nodes tightly to form a knowledge model.

8.3 Causal Reasoning

The clinical diagnosis in TCM is mainly based on the internal rules. In this chapter, we transform the diagnosis process into a layered causal graph.

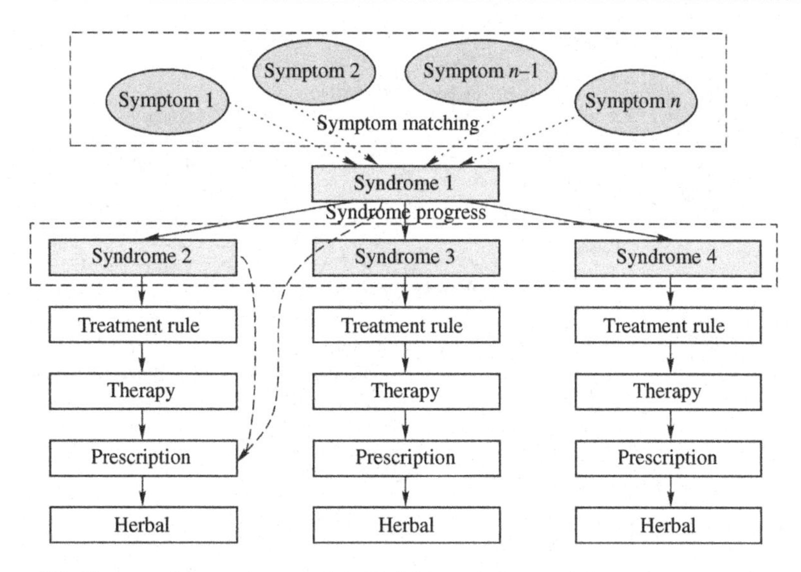

Figure 8.1 The causal reasoning graph in TCM knowledge model.

We define the layered causal graph based on the fact that TCM diagnosis mainly focuses on the syndrome. In a causal graph, there are nodes as terminology base and causal links. Thus, formally a Causal Graph $G = \{V, E\}$, is defined as:

- $V \subseteq R$, where $R = \{$Symptom, Syndrome, Treatment Rule, Therapy, Prescription, Herbal Medicine$\}$.
- $E \subseteq U$, where $U = \{\rightarrow, \dashrightarrow, \rightsquigarrow\}$.
  - $X \rightarrow Y$ represents that X has a *Direct Causation* relationship with Y meaning that X is a direct cause of Y independently.
  - $X \dashrightarrow Y$ represents that X has a *Logic Causation* relationship with Y as combined causation with logical operators, which exists from symptoms to syndrome.
  - $X \rightsquigarrow Y$ represents that X has a *Weighted Causation* relationship with Y that X plays a partial causation in Y, which exists from previous layers to prescriptions.
- A *Rule Pattern* is defined as a direct path from X to Z or a logical path from a set of X to Z, which means nodes connected by links can be defined as a rule.

The basic idea of Figure 8.1 is that a certain set of symptoms identifies a certain syndrome or disease (logic causation), and some syndrome or disease leads to the corresponding rule of treatment and therapy (direct causation). However, the final prescription will be decided from the symptoms, the developing syndromes, and the therapy together (weighted causation).

Generally, causal reasoning makes the TCM diagnosis process into a structural graph and provides different layers of the graph with suitable algorithms.

8.4 Evaluation

In this section, we discuss the experimental evaluation of our model. In total, we designed 821 classes, 26 object properties, 134 class assertion axioms, and

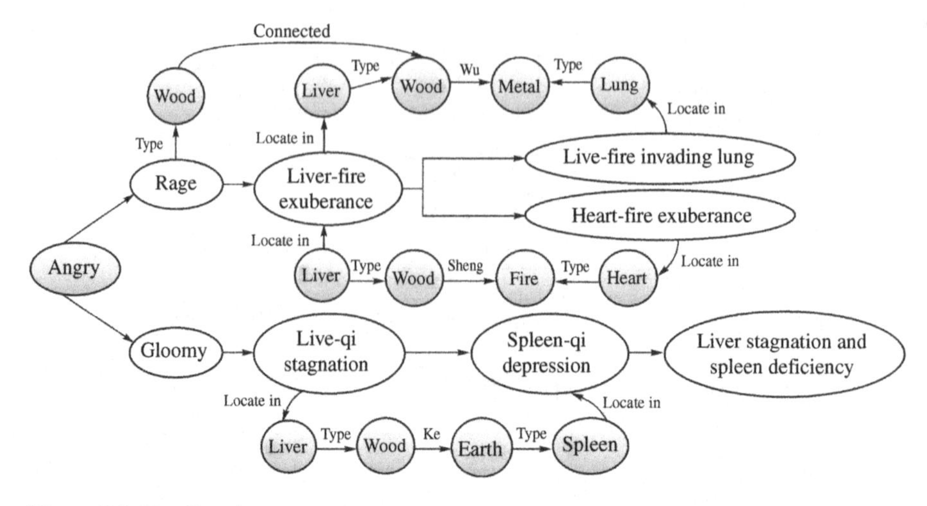

Figure 8.2 Use Case 1.

78 object property axioms in our ontology. Because our causal knowledge model is derived from TCM diagnostic principles, we evaluated the process using two typical medical use cases in TCM.

Use Case 1 (as shown in Figure 8.2*)*
Input: Angry (single symptom).
Output: A causal graph generated by user input as follows.
Description: When our system gets the input, it will search the RDF graph starting with the input node "angry," which means searching reachable nodes through property chains and predefined rules in the generated graph. All the white nodes are the final displayed results. The gray ones are the latent causal links upon each edge, which can be displayed when users click the edge.
Use Case 2 (as shown in Figure 8.3*)*
Input: A set of symptoms.
Output: One or multiple possible syndromes diagnosed from the input symptoms.
Description: When a user submits a set of symptoms, our approach searches possible corresponding syndromes in our knowledge base and outputs it to the user.

The evaluation results are represented in Chinese in our system, so we depicted them in use cases above using alternative graphs. As we described, Use Case 1 shows the important principles in the TCM philosophy, i.e., the action cycles between the basic five elements, based on which we conduct knowledge learning and knowledge mining. Our TCM knowledge model gets a satisfying result in TCM knowledge representation. It also has great potential in knowledge mining.

8.5 Conclusions

In this chapter, we present a causal knowledge modeling method that can be applied to the TCM diagnosis process. The principal objective of TCM knowledge

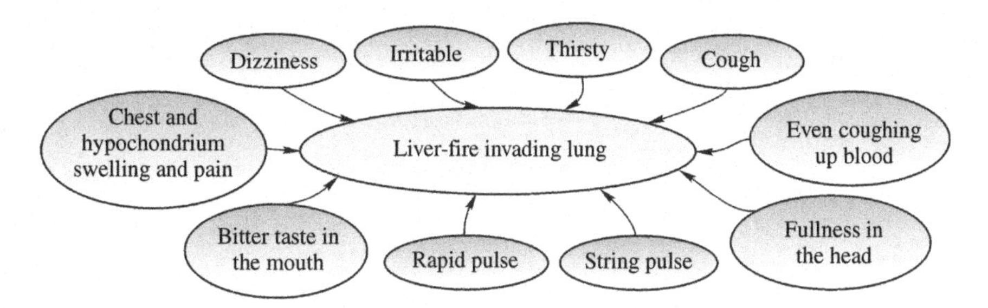

Figure 8.3 Use Case 2.

modeling is to figure out a formal method to represent Chinese medicine knowledge.

We build a TCM causal knowledge model based on the belief that the underlying causal relations inside TCM ontology can be represented using OWL 2. We defined seven ontology classes and corresponding subclasses based on Five Phase Theory. All the properties are defined according to the key activities between key concepts in the diagnosis. The relationship between symptoms and diseases or syndromes is the focus in determining the disease. However, the relationship is not a pure one-to-one relationship, and there are many uncertainties. For this reason, we viewed TCM knowledge as a layered causal graph and, in particular, viewed sets of symptoms as a whole, and our algorithms for the causal graph involve symptom matching and syndrome progress. Our causal knowledge reasoning, which is a method of integrating defined ontology with predefined rules to compose a causal graph, can clearly demonstrate the process of TCM diagnosis and show the potential to perform knowledge mining.

References

[1] Berners-Lee T, Hendler J, Lassila O. The Semantic Web. Sci Am 2001.
[2] Manola F, Miller E. RDF primer, <http://www.w3.org/TR/rdf-primer/>; 2004.
[3] Van Harmelen F, Bechhofer S, Hendler J, et al. OWL Web ontology language reference, <http://www.w3.org/TR/owl-ref/>; 2002.
[4] EB/OL: Wikipedia-Wu Xing, <http://en.wikipedia.org/wiki/Wu_Xing/>; 2009.
[5] Cheng Y, Wang Y, Wang X. A causal relationship discovery-based approach to identifying active components of herbal medicine. Comput Biol Chem 2006;30:148–54.
[6] Huang M, Chen M. Integrated design of the intelligent web-based Chinese Medical Diagnostic System (CMDS)—systematic development for digestive health. Expert Syst Appl 2007;32:658–73.
[7] Wang X, Qu H, Liu P, et al. A self-learning expert system for diagnosis in traditional Chinese medicine. Expert Syst Appl 2006;26:557–66.

9 Dynamic Subontology Evolution for Traditional Chinese Medicine Web Ontology[1]

9.1 Introduction

Traditional Chinese medicine (TCM) is one of the world's oldest medical systems with a time-honored history of several thousand years. TCM has made brilliant achievements and contributed significantly to the health of the Chinese people and the development of the world's medical science [1]. Currently, as a kind of wealth inherited from our ancestors, TCM has been studied and applied by scientists and doctors in many countries all over the world, especially in Asian countries such as Japan and Korea (also known as Traditional Medicine in those countries). Recent advances in the Web and information technologies with the increasing decentralization of organizational structures have resulted in a massive amount of TCM data distributed among the literature, clinical records, experimental data, databases, and so on [2]. There are many independently developed specialized data resources (medical formula databases, electronic medical record databases, clinical medicine databases, and so on) distributed among different TCM institutes. The data from those resources are potentially related to each other and they are necessary for TCM scientists to reuse them on a global scale. There is an increased emphasis on mashing up existing heterogeneous TCM data resources. However, information sharing also is impeded by the heterogeneity and distribution of the independently designed and maintained sources.

The use of ontologies [3,4] for the explication of implicit and hidden knowledge is a possible approach to encoding domain knowledge into reusable format and overcome the problem of semantic heterogeneity of data resources in different disciplines. The recent advent of the Semantic Web [5] has facilitated the incorporation of various large-scale online ontologies in the disciplines of biology and medicine, such as unified medical language system (UMLS) [6] for integrating biomedical terminology, gene ontology [7] for gene products, and MGED ontology [8] for microarray experiments. One important potential benefit of those activities is the bridging of the gap that exists between basic biological research and medical applications [9]. Therefore,

[1] Reprinted from Mao Y, Wu Z, Tian W, Jiang X, Cheung WK. Dynamic sub-ontology evolution for traditional Chinese medicine web ontology. *J Biomed Inform* 2008;41:790–805, © 2008 Elsevier Inc., with permission from Elsevier.

Modern Computational Approaches To Traditional Chinese Medicine. DOI: http://dx.doi.org/10.1016/B978-0-12-398510-1.00009-1

a unified and comprehensive TCM domain ontology is required to support TCM research, especially for TCM data resources mash-up. However, including large-scale ontologies in their complete form could incur an unnecessarily high storage and maintenance cost. For instance, as the size of an ontology scales up, the performance of the associated ontology-based systems will quickly degrade as frequent search and access of concepts in the large ontology are then required.

One possible approach for addressing the problem is to index the ontology to support efficient concept search and access. For example, the Semantic Web for History (SWHi) project [10] uses Lucene [11], a full-featured text search engine library to index ontologies. However, this approach achieves efficient search at the expense of a high storage requirement for the index, which is especially unfavorable when the ontology is very large scale. In this chapter, we take a different approach. By assuming that the problems of interest to a particular user group typically do not need the complete ontology but just a portion of it, we propose to incorporate into ontology-based systems the capabilities to (1) extract context-specific subontologies from the complete Web ontology, (2) cache them so as to be reused for subsequent tasks, and (3) evolve them to make the caching process adaptive, and thus more cost effective. In particular, a genetic algorithm (GA) is derived as the evolution mechanism for optimizing the subontologies' quality for dynamic knowledge reuse.

The remainder of this chapter is organized as follows: Section 9.2 presents a large-scale TCM Web ontology. Section 9.3 gives a formal definition for subontology, as well as a set of basic operations for manipulating subontologies. Section 9.4 depicts the ontology cache for knowledge reuse. Section 9.5 presents a subontology evolution approach based on the GA. Section 9.6 evaluates the approach with the TCM Web ontology. Section 9.7 gives an overview of the existing related work. Finally, we conclude this chapter with a look at future research issues in Section 9.8.

9.2 TCM Domain Ontology

In collaboration with the China Academy of Traditional Chinese Medicine (CATCM), we initiated the Traditional Chinese Medicine Language System (TCMLS) project in 2001, which is a unified TCM ontology [12] funded by China Ministry of Science and Technology to support terminology standardization, knowledge acquisition, and data integration in TCM. The ontology has become large enough to cover different aspects of the TCM discipline and is used to support semantic e-Science for TCM. As far as we know, TCMLS has been the world's largest TCM domain ontology.

9.2.1 Ontology Framework

As TCM evolves from generation to generation, a massive amount of concepts and relationships emerges from different sources; however, there is no unified terminology in the field of TCM. There have existed several controlled vocabularies and many special terminology lexicons in TCM, but combining and synchronizing individual versions of existing medical terminology vocabularies into a unified terminology system is still a tough issue, because of the heterogeneity and indistinctness in different lexicons. Besides, the content of the TCM ontology is acquired and generated dynamically from various knowledge sources in the process of development. We cannot determine how many classes (concepts), instances (individuals), and their relations (roles) will be generated on earth. Therefore, the development of the TCM ontology should conform to an incremental mechanism, which is adopted by many large-scale ontologies like Gene Ontology. We set up an upper-level unified framework (see Figure 9.1) for modeling the TCM ontology.

In order to feature the domain characteristics of TCM, we divide the whole TCM domain into several subdomains (see Table 9.1). A relatively independent category is corresponding to a relatively closed subdomain, compared with the TCM ontology corresponding to the whole TCM domain. In the current TCM ontology, there are 12 major categories, with each corresponding to a subdomain (Basic Theory of TCM, Formula of Herbal Medicine, Acupuncture, and so on) of TCM. The characterization of content of each category is listed in Table 9.1.

Considering medical concepts and their relationships from the perspective of the TCM discipline, we define the knowledge system of the TCM ontology by two components: concept system and semantic system. The semantic system concerns the basic semantic type and semantic relationship of the content class, which are ontology classes themselves (see Figure 9.2). According to the time, function space, entity, and class attributes of TCM domain knowledge, we have defined 59 kinds of semantic relationships between content classes and about 104 kinds of TCM semantic types, plus all the semantic types from UMLS.

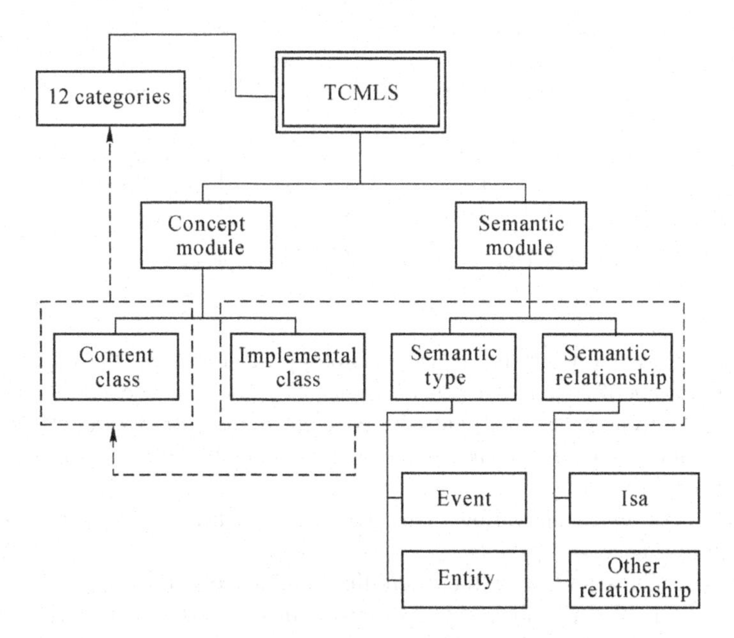

Figure 9.1 The upper-level framework of TCM ontology.

Table 9.1 The 15 Categories defined in the TCM Ontology Corresponding to the Subdomains of TCM

Categories	Concepts	Individuals
Pathogeny and pathogenesis and diagnosis	65	4,939
Geography	948	86
Pharmacology of TCM medical formulas	158	11,677
Disease	308	6,516
Humanities	1680	10,448
Health care management	303	1,300
Medicinal plants and animals	719	14,804
Informatics and philology	216	21,089
Prevention and life cultivation	163	69
Acupuncture	716	1,647
Therapeutic principles and treatments	108	1,997
Traditional Chinese drugs	1572	3,624
The basic theory of TCM	3313	1,725
The doctrines of TCM and relevant sciences	1352	219
Natural science and physics	217	240

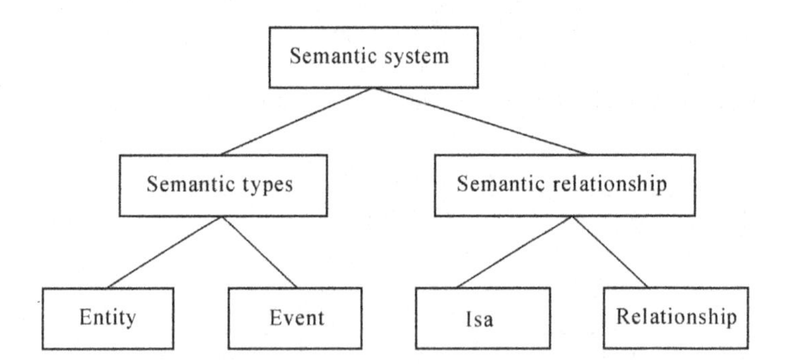

Figure 9.2 The semantic system framework of the TCM ontology.

We unify class structure in the TCM ontology and formulize the key properties and relationships of content class as implemental classes. The concept system contains a content class that represents the concrete domain knowledge of the TCM discipline and four kinds of basic implemental classes to define the instances of a content class.

Name Class represents various name terms (e.g., alias, English name, or synonym) of an instance.

Definition Class represents the scientific definition of an instance. Explanation Class represents the additional explanations to instance definition. Relation Class represents the relationship between two instances.

The implemental classes and the semantic system are used to compose the content class. A class has literal property and object property in ontology. The range of a literal property is a literal or string, while the range of an object property is a class. A content class in the TCM ontology has five major object properties, each related to a class. We can create the new instances of implemental classes incrementally to compose the instances of the content class and set up the relationships between instances. For example, the relation class has two major object properties: one ranges the semantic relationship and the other one ranges the content class, which means content classes are related to each other through semantic relationships.

In this way, all content classes in the TCM ontology have a unified knowledge structure while different instances of content class have contents and relationships of their own. Various content classes cluster into 12 categories to formulate the domain knowledge of the TCM discipline. Ontology developers can create new subclasses in the high-level framework and populate new instances on-demand, without modifying class structure.

9.2.2 User Interface

In order to facilitate using the TCM ontology development, we have built a Web-based TCM ontology editor (see Figure 9.3). It runs on the server-side and publishes large-scale TCM ontologies to users through Web services. The editor allows

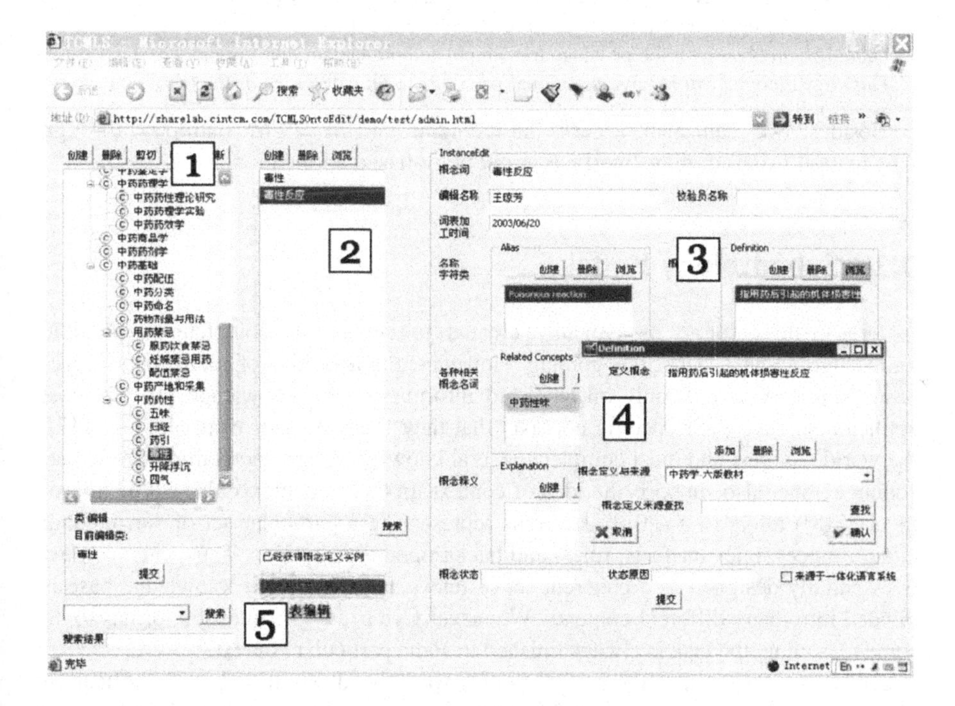

Figure 9.3 The default user interface for the TCM ontology editor.

users to edit and explore ontology online. No special clients are required and users can browse and edit the TCM ontology anywhere with their Web browsers. The editor supports cooperative online ontology editing, incorporates a database backend for large-scale ontology storage, and interacts with several popular ontology formats like RDF [13], OWL [14], and so on. The editor provides a tree-based view for classes and a form-based view for instances.

Figure 9.3 depicts the standard user interface of the editor. The user interface incorporates an open-source AJAX framework [15], which enables immediate data update without refreshing Web pages in Internet browsers. The interface is divided into a set of panels. Figure 9.3 has three panels visible. The class panel shows the class hierarchy (1), the instance panel shows a list of instances (2), and the property panel shows the details of a selected instance (3). The property panel consists of several slots, with each corresponding to a class property. For a specific instance, users can specify its property value by inputting either literals or related instances in a pop-up panel from slots (4). Also, the editor provides some additional functions like searching a class or instance in the scope of the ontology, counting the number of classes or instances and so on (5), to facilitate development.

We employ a layered privilege mechanism in the ontology editor, and users who play different roles in the ontology hold different privileges. There are mainly four kinds of roles in the TCM ontology: reader, editor, checker, and administrator.

- *Ontology readers* are able to browse all the contents of the whole ontology.
- *Ontology editors* are able to input, modify, and delete instances within a category but have no privilege to manipulate the classes of the category.
- *Ontology checkers* own the privilege to manipulate both classes and instances in a category.
- *Ontology administrators* have the global privilege to all categories of the whole ontology.

Besides, we can also access the contents of the TCM ontology through Application Programming Interface in ontology-based applications.

9.3 Subontology Model

A large-scale ontology is typically created to contain as complete as possible knowledge about a particular domain. However, the activities of many knowledge-based applications rely only on localized information and knowledge [16] as most applications have their specific contexts that they focus on. For example, Cyc [17], the world's largest and most complete general knowledge base and commonsense reasoning engine, also supports the idea of context. In Cyc, a context (also called microtheory [18]) refers to a set of assertions representing a particular set of surrounding circumstances, relevant facts, rules, and background assumptions. Thus, each context is essentially designed as a coherent set of relevant material. The knowledge base is divided into many different contexts. Whenever Cyc is asked a question, or has to do some reasoning, the task is always finished in some particular context.

Our conjecture is that the activities of a specialized ontology-based application need only some specific aspects of a complete ontology. Small and context-specific portions of an ontology are often needed in semantic-based applications, especially those with limited capacity such as embedded systems or mobile agents. For example, a TCM ontology contains domain knowledge on different aspects of TCM. Figure 9.4 depicts a part of the content about the Science of Traditional Chinese drugs (TCD) from the TCM ontology. If we want to build a semantic-based web site specialized on a well-defined category of knowledge (e.g., herbal medicine), we can easily identify and extract the corresponding portion of knowledge from the TCM ontology. All applications supporting that web site then need only process that portion of ontology, instead of the complete one. DartGrid [19], initiated in 2002 by Zhejiang University, is a semantic-based platform for sharing TCM knowledge and information for CATCM. In DartGrid, a large-scale TCM ontology is used to integrate TCM databases by the semantic mapping mechanism. Whenever the ontology is required in establishing semantic mapping, the system has to access the TCM ontology, which is large scale and therefore leads to a high cost. In fact, DartGrid uses only portions of the TCM ontology when integrating databases. Figure 9.5 depicts a sample scenario for integrating TCD databases by the ontology in Figure 9.4. Given several TCD databases, the TCM ontology is used to integrate them by creating mapping between the schemata of tables and the ontology. Only the part with a dashed line in Figure 9.4 is used for database integration, while other contents are not used at all.

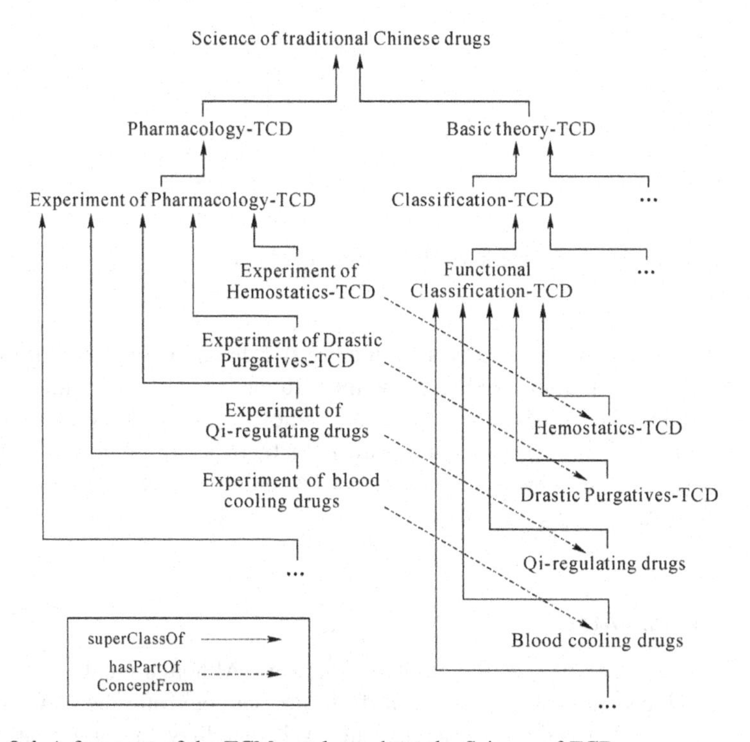

Figure 9.4 A fragment of the TCM ontology about the Science of TCD.

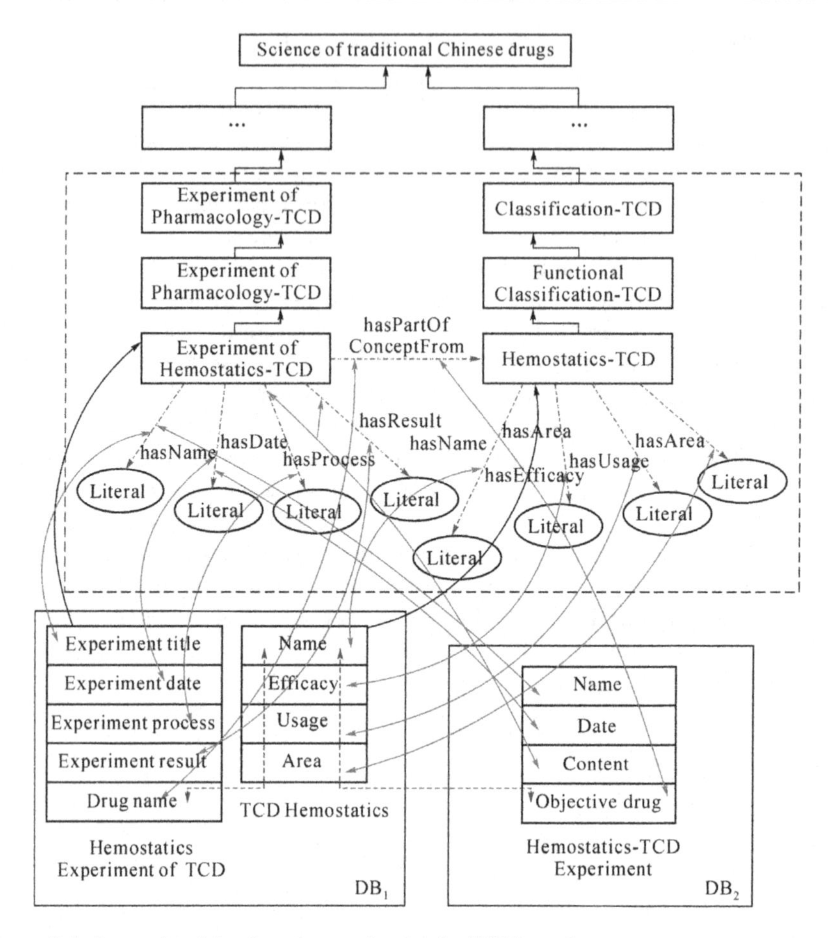

Figure 9.5 A sample of database integration by the TCM ontology.

This calls for the ability to extract from a large-scale ontology those corresponding context-specific portions and to allow them to evolve (iteratively throughout a sequence of knowledge-based activities) to achieve specialization in their own areas of concern. We represent and extract context-specific subontologies [20] from the complete ontology as an ontology cache and allow them to evolve according to past experience to gain optimality.

9.3.1 Preliminaries

In this work, we use OWL as the ontology language. Although the TCM ontology was not developed in OWL, we can transform the ontology into the OWL format easily by the ontology editor. Generally, an OWL Ontology (or simply an ontology) O is a pair $\langle T, A\rangle$, where T and A are its TBox and ABox, respectively, in the description logic (DL) [21]. The TBox introduces the terminology like the vocabulary of an application domain while the ABox contains assertions about named individuals in terms of the vocabulary.

In OWL ontology, there is no direct definition about a link; however, we can interpret property restrictions as cross-links between different concepts.

Definition 9.1. (Cross-Link). Given an ontology $O = \langle T, A \rangle$, a cross-link, L, between two concepts is one of the following OWL property restrictions [14]: *range/domain, someValuesFrom, allValuesFrom.*

Cross-links are directional. To a symmetric relation between two concepts, we can interpret it as two cross-links from both directions of the two concepts. However, the TCM ontology does not contain such kinds of relations. Therefore, we do not deal with them in this chapter.

Definition 9.2. (Triple). Given an ontology $O = \langle T, A \rangle$, a triple, t, is represented as $\langle c_1, r, c_2 \rangle$, where c_1, $c_2 \in T$ are concepts and r denotes either the subsumption relation or a kind of cross-link.

An OWL ontology can be treated as a set of semantically related triples. By considering concepts as nodes and relations as arcs, an ontology can be represented as a graph. The number of arcs in the graph defines the number of triples in the ontology. Note that the graph is directed as the relations between two concepts are directional. Figure 9.4 illustrates a sample ontology in the form of a graph. The ontology is selected from the TCM ontology.

Definition 9.3. (Relation Path). Given a set of triples $T = \{t_i | t_i = \langle A_i, R_i, B_i \rangle, i = 1, 2, \ldots, n\}$, and two triples $t_s = \langle A_s, R_s, B_s \rangle$, $t_r = \langle A_r, R_r, B_r \rangle$, t_i, $t_j \in T$, if there exists $m \leq n - 2$ and $t_s \cdot B_s \equiv t_i \cdot A_i$, $t_i \cdot B_i \equiv t_{i+1} \cdot A_{i+1}, \ldots, t_{i+m-1} \cdot B_{i+m-1} \equiv t_r \cdot A_r$, then we say there is a relation path from t_s to t_r.

9.3.2 Subontology Definition

In the following, we give a formal definition of subontology to facilitate the discussion in the sequel.

Definition 9.4. (Subontology). Given an ontology $O = \langle T, A \rangle$, a subontology (SubO for short), B, is represented as a tuple $\langle B_c, B_t, B_a, O \rangle$, where $\langle B_c \subseteq T, B_a \subseteq A \rangle$, B_a A_i denotes the local knowledge base of the SubO, and B_c serves as the SubO's context description which contains some concepts $\{c_i \in B_t\}$, which are by definition required to be pair-wise interrelated in B_t.

Given B_c, one can derive the SubO by searching the ontology O for related concepts and storing them as B_t and B_a. Given a context $B_c = \{$Hemostatics-TCD, Drastic Purgatives-TCD, Classification-TCD$\}$, we can obtain the SubO as shown in

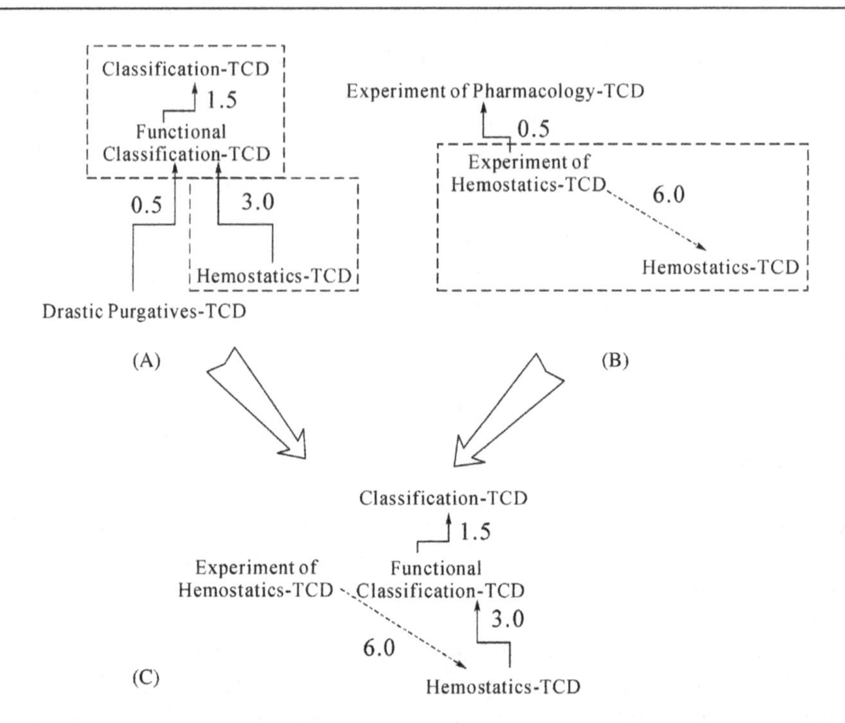

Figure 9.6 Sample SubOs from the TCM ontology in an ontology cache.

the dashed line of Figure 9.6A. Note that B_a is a set of individuals from the ABox of the source ontology, which is used to make a SubO complete semantically. However, we focus on searching and reasoning TBox in this work, and B_a is not actually used.

To contrast with the microtheory concept in Cyc, we extract SubOs dynamically from source ontology and allow them to evolve in the process of knowledge-based activities like reasoning (to be discussed later), while microtheories in Cyc are static and an invariant division of the knowledge base. Note the proposed evolution itself will not produce new knowledge, but it will improve the local knowledge structure of ontology-based systems which can then reuse SubOs in a more dynamic and adaptive way. Compared with SubO, the complete domain ontology serves as a static and invariant resource for the ontology-based or semantic-driven systems. Revisions of domain ontologies require owner privilege and domain experts' intervention, and thus it is impractical to directly modify and update source ontologies based on diverse needs of different applications. How can their local repositories of semantics be extracted from domain ontologies in the form of local SubOs as described in this section?

9.3.3 Subontology Operators

In this section, we describe a set of operators needed for manipulating context-specific SubOs in an ontology. They include extract, store, compare, retrieve, and merge.

9.3.3.1 Extract

Given an ontology O, a concept set $C_s = \{c_i\}$, the extract operator's input can be represented as a triple $\langle C_s, n, O\rangle$, where n is the traversal depth. It will return a SubO B.

The procedure of the extraction is as follows:

1. Randomly select a concept c in C_s, remove c from C_s, and add c to a set B_c.
2. Perform breadth-first traversal through cross-links. If another concept (c') in C is reached in traversal, allocate the triples along the traversed path from c to c_0 into a result set B_t. Remove c_0 from C_s and add c_0 to B_c. Terminate the traversal up to a depth of n.
3. Repeat steps 1 and 2 until C_s is empty.
4. Get the set of individuals B_a, for all the concepts in B_t.
5. The corresponding SubO $B = \langle B_c, B_t, B_a, O\rangle$.

9.3.3.2 Store

Given a SubO $B = \langle B_c, B_t, B_a, O\rangle$, a SubO repository R, the store operator's input is a pair $\langle B, R\rangle$. We can implement the ontology repository in a relational database. Then B_c, B_t, and B_a are stored in three fields, respectively, as well as an additional field pointing to O.

9.3.3.3 Compare

Given an ontology O, two SubOs B_1 and B_2 to be compared, the compare operator takes a triple $\langle B_1, B_2, O\rangle$ as its input. The operation will return a similarity degree *sim*.

The discrepancy is computed based on their edit distance or the Levenshtein distance [22]. There are two possible strategies for computing the edit distance. One is based on the edit distance of two context descriptors B_{c_1} and B, where the distance is defined as the number of changes required to turn one concept set into another. The other is based on the edit distance of the knowledge contents of the two SubOs $K_1 = \langle B_{t_1}, B_{a_1}\rangle$ and $K_2 = \langle B_{t_2}, B_{a_2}\rangle$. Given either form of edit distance, the size of the SubO's context descriptor, or knowledge content being not invariant makes direct distance comparison difficult. Instead, we use relevant edit distance instead of the conventional edit distance directly.

Definition 9.5. (Relevant Edit Distance). Given two SubOs $B_1 = \langle B_{c_1}, B_{t_1}, B_{a_1}, O\rangle$, $B_2 = \langle B_{c_2}, B_{t_2}, B_{a_2}, O\rangle$ and the edit distance ED ( ), the relevant edit distance (RED) between B_1 and B_2 is computed as follows:

$$\mathrm{RED}(B_1, B_2) = \mathrm{ED}(K_1, K_2)/|K_1| \text{ or } \mathrm{ED}(B_{c_1}, B_{c_2})/|B_{c_1}|$$
$$\text{where } K_1 = \langle B_{t_1}, B_{a_1}\rangle \text{ and } K_2 = \langle B_{t_2}, B_{a_2}\rangle.$$

If we want to get a more accurate semantic similarity between two SubOs, we should make use of the semantics of the ontology. For example, Mena et al. [23] proposed an approach to estimating the information loss by substituting terms from one ontology with terms from another ontology. When we compare two SubOs, there is also information loss between two SubOs. However, our approach is simpler and easier to implement, although it has information loss for semantic-based applications when two SubOs are compared.

9.3.3.4 Retrieve

Given a context description D, a SubO repository R, the retrieve operation is a triple $\langle D, R \rangle$. The operation will return a SubO B.

A best-matched SubO from a SubO repository can be retrieved by computing the RED between D and the concept sets of the SubOs in R. Here, being best matched means that not only the retrieved SubO has the highest similarity but also the similarity is higher than a threshold.

9.3.3.5 Merge

Given an ontology O and two SubOs, $B_1 = \langle B_{c_1}, B_{t_1}, B_{a_1}, O \rangle$ and $B_2 = \langle B_{c_2}, B_{t_2}, B_{a_2}, O \rangle$ to be merged, the merge operator takes a triple $\langle B, O \rangle$ as its input and returns a SubO B as its output.

The aim of merging is to combine two SubOs from the source ontology together. As the two SubOs are extracted from the same source ontology, the merging operation is not very difficult. The merging operation is similar to a union to some extent. The procedure of merging is as follows:

1. Merge B_{c_1} and B_{c_2} into a new set B_c.
2. Merge $\langle B_{t_1}, B_{a_1} \rangle$ and $\langle B_{t_2}, B_{a_2} \rangle$ into a new pair $\langle B_t, B_a \rangle$.
3. Check the connectivity of concepts in B_t, if $\exists c_i, c_j \in B_t$, and if there exists no path from c_i to c_j in B_t, traverse from c_i to c_j in O and add the triples in the path to B_t.
4. Get the set of individuals for the new concepts in B_t and add them to B_a.
5. The corresponding SubO $B = \langle B_c, B_t, B_a, O \rangle$.

9.4 Ontology Cache for Knowledge Reuse

We introduce the subontologies as an ontology cache to support a knowledge search in ontology-based systems.

9.4.1 Reusing Subontologies as Ontology Cache

As mentioned before, context-specific portions of the TCM ontology represented as SubOs can serve as an ontology cache to support the knowledge-based activities of ontology-based systems. The concept of an ontology cache draws inspiration from the memory-caching mechanism to support efficient reuse of SubOs. First emerging from data processing, caches work well because of a principle known as locality of reference. An ontology cache refers to a local repository for retrieved portions of ontology (in terms of SubOs) for future reuse. Taking SubOs as cacheable objects, an ontology cache stores SubOs as cache blocks for ontology reuse. When an ontology-based system lacks the knowledge to perform reasoning tasks, it needs to access related ontologies. If the ontology-based system extracts the required SubOs and keeps the most active ones in an ontology cache, the overall speed for reasoning is then optimized.

Therefore, we can complement the database integration of DartGrid with the SubO mechanism. Caching the SubOs from the TCM ontology, which are frequently used in database integration, can improve the performance of DartGrid. When we want to integrate new databases in DartGrid, the system first looks for a SubO in its ontology cache to provide formal semantics rather than accessing the TCM ontology directly. For example, there are three SubOs that are all extracted from the ontology in Figure 9.4 in the ontology cache. Given the databases in Figure 9.5, the system can search the ontology cache in Figure 9.6 to find SubO (c) for integrating the databases.

SubOs are extracted from large-scale ontology and cached for future reuse. If the size of an extracted SubO is still large, it will lead to the low efficiency of ontology-based systems in ontology reuse. There are several methods to deal with the problem mentioned above. First, we can use the depth of extraction to constrain the size of a subontology. Second, subontologies are stored in an ontology cache, so the cache space also limits the size of SubOs. If a SubO is too large, for example, over a threshold size, it cannot be accepted by the cache.

Caches can be divided into tuple caching, page caching, and semantic caching [24] according to the granularity. Tuple caching is based on individual tuples, page caching is based on pages which are statically defined groups of tuples, and semantic caching uses dynamically defined groups of triples (also called semantic regions or semantic segments). According to Ref. [24], both page and semantic caching reduce overhead by aggregating information about groups of tuples. However, semantic caching provides more flexibility, allowing the grouping to be adjusted to the needs of queries. Therefore, we use SubOs instead of triples as basic cache blocks for the ontology cache. SubO caching can be treated as a kind of semantic caching for ontology triples.

The major tasks for this work can be summarized as follows:

1. Search knowledge contents in an ontology cache for the TCM ontology.
2. Optimize the knowledge structure of SubOs in the ontology cache for better performance.

9.4.2 Knowledge Search with Ontology Cache

One of the most important tasks for the TCM ontology is to support reasoning in ontology-based systems. In DL settings, not all entailments are equally valued. Grau et al. [25] list three main reasoning tasks on which DL systems have traditionally focused:

1. Atomic concept satisfiability (A) determines whether an atomic concept in the ontology is satisfiable;

2. Classification (C) computes the subsumption partial ordering of all the atomic concepts in the ontology;
3. Instantiation and retrieval (I) determine whether an individual is an instance of an atomic concept and retrieve all the instances of an atomic concept, respectively.

In addition, a SubO contains a number of relation paths and there is a certain relationship between two concepts in a path. However, the relationship is not handled by all the tasks above. Therefore, we add another reasoning task called indirect relationship (R), which determines whether there is a relation path between two concepts. For example, there is a relation path from the concept Experiment of Hemostatics-TCD to the concept Classification-TCD as we can see in Figure 9.6C, so Experiment of Hemostatics-TCD has an indirect relation with Classification-TCD.

In this chapter, we call a reasoning task a question, which is defined as follows:

Definition 9.6. (Question). Given an ontology $O = \langle T, A\rangle$, a question, q, is a triple $\langle Q_b, Q_c, Q_o\rangle$, where Q_o is an ontology, $Q = \{\text{ontology concept } c_i\}$, where $\forall c_i \in Q_o$, and Q_c is a condition tag that denotes one of the reasoning tasks above.

If we want to transform a natural user description to an explicit semantic description of the question automatically, it will fall into the fields of human computer interaction and natural language processing. Therefore, we simplify the questions from users or applications as a list of concepts and conditions as Definition 9.6 is convenient for simulation.

The domain knowledge involved in SubOs is reused to support reasoning tasks like a question. We try to answer questions by searching knowledge among a collection of SubOs in an ontology cache. For example, there is a question $Q =$ {Experiment of Hemostatics-TCD, Experiment of Pharmacology-TCD}, C, O_i, where Otcm refers to the TCM ontology. The meaning of Q is to determine the subsumption between Experiment of Hemostatics-TCD and Experiment of Pharmacology-TCD. Assume there are three SubOs in an ontology cache (see Figure 9.6). Then we find out SubO (b) in the cache and give a positive answer to Q according to the semantics of SubO (b).

In an ontology without full negation, the reasoning tasks mentioned above can be reduced to unsatisfiability. Therefore, we try to answer a question for atomic concept satisfiability by reusing a collection of SubOs. Assume the question is in terms of h, D, A, O_i, and then we have to check the satisfiability of the concept description D in order to answer the question. Algorithm 9.1 is as follows:

Algorithm 9.1 Check SubO Satisfiability

```
Input: a collection of SubOs S, a concept description D, the source ontology O
Output: a boolean value ret
Comments: check the satisfiability of a concept respect to a collection of
SubOs
for each SubO Si in S
```

```
    if Si contains the terms of D then
      ret = tableau(D, Si)//using the tableau algorithm to check the
satisfiability of D with respect to Si
    end if
    return ret
end loop
//there is no SubO that contains the terms of C
ret = tableau(D, O )//using the tableau algorithm to check the satisfiability
of D
return ret
```

In Algorithm 9.1, if no SubO can be selected as the knowledge base for the concept, we can apply the transformation rules in the tableau algorithm [26,27] to check the satisfiability, with respect to the whole source ontology. Otherwise, we apply the tableau algorithm to check the satisfiability, with a SubO as the local knowledge base for reasoning. If we take a SubO as the knowledge base, then the process of reasoning is almost the same as that in the ontology. The major difference is that the reasoning in a SubO is semi-deterministic. When a concept is unsatisfiable in a SubO, it is unsatisfiable in the source ontology, too. However, if a concept is satisfiable in a SubO, it is not always satisfiable in the source ontology. For example, the source ontology contains the concept description $\neg A \cap B$. We extract a SubO from the ontology and $\neg A \cap B$ is not involved in the SubO. When we check the satisfiability of the concept description $A \cap \neg B$ based on the SubO, we cannot get an unsatisfiability. However, $A \cap \neg B$ is obviously inconsistent with the source ontology. Although the reasoning in SubO is not complete, we can still use SubOs to improve the performance of ontology-based systems in some aspects.

There are two top-level concepts (the direct subconcept of the root concept) in the TCM ontology Otcm:

- *Chinese Materia Medica (CMM)*: This, also known as herbal medicine in TCM, contains the natural medicinal materials, such as plants, animals, and minerals, used in TCM practice.
- *Formula of Herbal Medicine (FHM)*: This defines the basic notions such as benzoin tincture compound, rhubarb compound, and cinnabar compound, which are based on the theory of prescription of formula.

We can extract a SubO *B* that contains these two concepts and *B* is shown as the part with dashed lines in Figure 9.7. As we can see, *B* only contains a small part of knowledge from Otcm.

There are two instances of CMM: *Glycyrrhiza* (g), which is an instance of legume, and *Sargassum* (s), which is an instance of gulfweed. According to the basic theory of TCM formula, *Glycyrrhiza* and *Sargassum* cannot be used as components in the same formula. Therefore, we have the following axiom in Otcm:

$$F_g \cap F_s \subseteq \perp$$

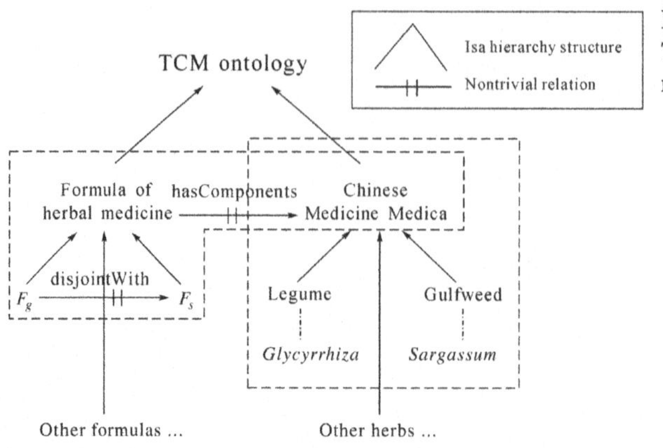

Figure 9.7 SubOs from the TCM ontology to support reasoning.

Where the concept F_g denotes the collection of formulas that contains *Glycyrrhiza* in its components and F_s denotes the collection of formulas that contains *Sargassum* in its components.

Assume that a user or application forms an ontology concept $F_{new} \equiv F_1 \cup F_2$, in order to represent a new kind of formula, where the concept F_1 denotes a collection of formulas that contains *Glycyrrhiza* in its components and F_2 denotes a collection of formulas that contains *Sargassum* in its components. Therefore, we have the following axioms:

$$F_1 \subseteq F_g, \quad F_2 \subseteq F_s$$

This raises the problem of checking whether the concept description $F_{new} \equiv F_1 \cap F_2$ is satisfiable or not. The question is in terms of $\langle F_1 \cap F_2, A, O_{tcm} \rangle$. We can use the tableau algorithm to achieve this goal by reusing B. Assume F_{new} is also satisfiable and then $F_1 \cap F_2$ is also satisfiable.

$F_1 \cap F_2$	
F_1, F_2	
$F_1, F_2, F_1 \subseteq F_g, F_2 \subseteq F_s$	
$F_1, F_2, F_1 \subseteq F_g, F_2 \subseteq F_s$	
F_g, F_s	
$F_g, F_s, F_g \cap F_s \subseteq \perp$	
$F_g, F_s, \neg(F_g \cap F_s)$	
$F_g, F_s, \neg(F_g \cup \neg F_s)$	
$F_g, F_s, \neg F_g$	$F_g, F_s, \neg F_s$
Clash	Clash

Reasoning step by step, we get a contradiction finally. The concept description $F_{new} \equiv F_1 \cap F_2$ is unsatisfiable with respect to B. Moreover, it means the new formula is not valid for TCM.

When a collection of TCM-specific questions arrives, an ontology cache will try to answer each question by searching and reusing existing SubOs. If it can retrieve a SubO to answer the question, we say it is a cache hit, otherwise it is a miss. If a cache miss occurs, the cache will try to extract SubOs from the source ontology and the description of a question is used as the concept set for extracting a SubO. A newly extracted SubO will be placed into the cache if the cache is not full; otherwise, the cache will try to replace the existing SubOs with the new one.

9.4.3 On SubO Structural Optimality

When a SubO is used to answer a question, not all parts in the SubO are equally used. When a triple is used as a part of a proof for answering the question, it gets an incremental value. Therefore, different triples in a SubO will gain different values. We define every m ($m \geq 1$) question as a section. The cache value of a triple is updated in every section. The initial value (section 0) of a triple is zero. The cache value of a triple t is computed as follows:

$$cv_n(t) = \begin{cases} 0 & \text{if } n = 0 \\ \Delta cv_n(t) + \lambda \cdot cv_{n-1}(t) & \text{otherwise} \end{cases} \tag{9.1}$$

where n equals the number of section(s), k is the expired rate of the cache value t obtained in the last section, and $\Delta cv_n(t)$ is the hit number of t during section n.

Equation (9.1) takes into account both the past experience and recent performance for each triple. Then the cache value for a SubO, B in section k in an ontology cache is as follows:

$$cv_k(B) = \sum cv_k(t)/n \tag{9.2}$$

where n is the number of triples in B.

As the size of SubO is not fixed and there exists redundancy among SubOs, a SubO with a high value does not stand for a good SubO in an ontology cache. For example, there are two SubOs in an ontology cache in Figure 9.6. The cache value for SubO (a) is $(3.0 + 1.5 + 0.5)/3 = 1.67$, and the cache value for SubO (b) is $(6.0 + 0.5)/2 = 3.25$. It seems that SubO (b) is a good SubO in the ontology cache. However, if we connect SubO (a) and SubO (b) at Hemostatics-TCD and trim some triples with low cache value (<1.0), we can get a new SubO (see Figure 9.6C), the cache value of which is $(3.0 + 1.5 + 6.0)/3 = 3.5$. Therefore, we get an optimized SubO by connecting and trimming (a more complete evolution approach is discussed later). Therefore, we should go beyond using traditional cache replacement policy, in order to improve the performance of the ontology cache. We must optimize the overall structure of SubOs in the ontology cache. Such optimization can be achieved by some evolutionary computational algorithms.

9.5 Dynamic Subontology Evolution

The term Ontology Evolution in the area of ontology engineering is treated as a part of the ontology versioning mechanism that is analyzed in Ref. [28], which refers to access to data through different versions of an ontology. Noy and Klein [29] just combined ontology evolution and ontology versioning into a single concept defined as the ability to manage ontology changes and their effects by creating and maintaining different variants of the ontology. To contrast with ontology evolution, our concern with subontology evolution is inclined to evolve a highly dynamic update of the SubOs in a local repository (ontology cache) to support knowledge-based activities of ontology-based systems.

As a kind of evolutionary computation, the theoretical foundations of the GA were proposed originally by Holland [30] in the early 1970s. It applies some of the natural evolution principles like crossover, mutation, and survival of the fittest to optimization and learning. GA has been applied to many problems of optimization and classification [31,32]. We here adopt GA to solve the problem of structure optimization to support SubO caching.

9.5.1 Chromosome Representation

For GA, a chromosome is composed of a list of genes (allele) and a set of chromosomes grouped together as a population. Chromosomes encoding in GA is to map the problem space to the parameter space. For the problem of SubO evolution, it refers to how to represent SubOs in the ontology cache as chromosomes. It is necessary to consider the features of SubO itself when performing chromosome encoding. We take the SubOs in the ontology cache as the problem space. Export all triples of SubOs in the cache as a triple list. Each allele in a chromosome denotes a triple. There are several possible approaches to represent chromosomes for SubOs:

- *Concept-based binary encoding*: Export all concepts of SubOs in the ontology cache as a concept list. Each allele in a chromosome denotes a concept. Each concept appearing in a SubO will have the allele in the corresponding chromosome set to 1 or 0 otherwise. We do not encode roles into a chromosome in order to simplify the representation in this approach.
- *Triple-based binary encoding*: Export all triples of SubOs in the ontology cache as a triple list. Each allele in a chromosome denotes a triple. Similar to the first approach, each triple appearing in a SubO will have the allele in the corresponding chromosome set to 1 or 0 otherwise.
- *Triple-based nonbinary encoding*: Export all triples of SubOs in the ontology cache as a triple list. Each allele in a chromosome denotes a triple; however, the allele is a nonbinary string in terms of the concept-role-concept.

We illustrate the three different encoding approaches through the example in Figure 9.8. For the ontology cache in Figure 9.6, we can represent the SubOs as chromosomes by different encoding approaches. The chromosomes generated by three different encoding approaches are listed in Figure 9.8A–C. As there is at

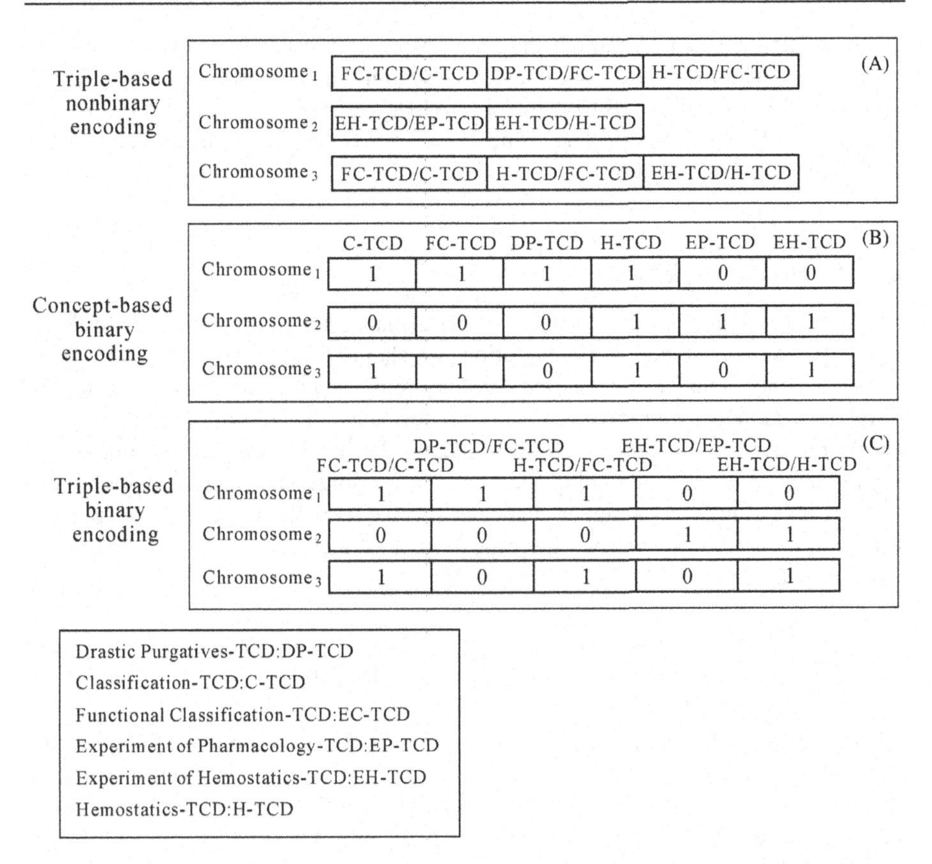

Figure 9.8 The results of the three different encoding approaches to the same set of SubOs.

most one role between two concepts, so role strings for each triple are omitted in Figure 9.8.

When the evolution terminates, we should decode chromosomes to SubOs. As concept-based binary encoding does not encode triple information in chromosomes, we have to reextract triples from the source ontology. A triple is represented as a binary character 0 or 1 in triple-based binary encoding encodes, so it cannot guarantee the connectivity of SubOs when we apply genetic operators like a crossover to chromosomes. In fact, both methods lose some information on encoding.

Therefore, we adopt triple-based nonbinary encoding in our evolution algorithm. A nonbinary alphabet in encoding is used. The concept names and role names of the SubOs in an ontology cache are used as the alphabet of encoding. The length of the chromosome is variable. When we encode a SubO as a chromosome, what we do is to mark each triple in the SubO as allele. The triple-based nonbinary encoding is to some extent like an index of the triples of a SubO.

The representation of SubOs does not destroy the constraints of concepts and the relations between concepts in the source ontology. It is only an abstract representation for GA. Although the chromosome representation does not encode everything from the source ontology, the relations and constraints are still preserved in the original SubOs of the cache. When we finish the evolution of GA and decode the chromosomes into new SubOs, we can get the relations and constraints back from the original ones.

9.5.2 Fitness Evaluation

The evaluation function is a measure of performance for SubOs. In order to determine an evaluation function for evolution, we should think of the criteria for measuring whether a SubO is better for ontology reuse or not. The major objective of SubO evolution is to improve the cache performance. Therefore, the fitness function in our evolution approach mainly evaluates the fitness of a chromosome by the cache value of the corresponding SubO. A chromosome, the SubO of which has a higher cache value, gets a higher chance to survive in evolution. Let P be a population with n chromosomes. Then the fitness value of a chromosome S in P is calculated as follows:

$$f(s) = \begin{cases} 0 & \text{if } B \text{ is disconnected} \\ cv(B)/\sum cv(B_1), \quad i = 1, 2, \ldots, n & \text{otherwise} \end{cases} \tag{9.3}$$

where B is the corresponding SubO for S and B_i is the corresponding SubO for the ith chromosome in P. In order to evaluate each chromosome, we need to compute the cache value of the corresponding SubO. For each chromosome, we calculate the cache value of its corresponding SubO obtained by Eqs. (9.1) and (9.2).

The overall fitness value of P is calculated as follows:

$$f(P) = \sum f(S_i)/n, \quad i = 1, 2, \ldots, n \tag{9.4}$$

According to Definition 9.3, the concepts in a SubO should be connected. However, we may get a disconnected SubO in evolution. In order to suppress the emergence of disconnected SubOs, we just give them a value of zero (fitness value should be non-negative). It makes sense that a connected SubO contains more information than a disconnected one and thus should be preserved in evolution.

9.5.3 Genetic Operators

After an initial population is generated, a GA carries out some genetic operators to generate offspring based on the initial population. Once a new generation is created, the genetic process is performed iteratively until an optimal result is found. Based on the chromosome representation and the fitness evaluation, we propose the genetic operators for SubO evolution. Selection, crossover, and mutation are the three most important and common operators of a GA. GA operators are triple-based operations to maintain semantic connectivity in SubO. We can implement the operators based on the SubO operations.

9.5.3.1 Selection

The selection operator uses the roulette wheel method, in which the selection probability of the chromosome is defined below:

$$P_i = f'(S_i)/f_{\text{sum}}, \quad i = 1, 2, \ldots, n \tag{9.5}$$

where n is the number of chromosomes in the population; $f'(S_i) = f(S_i)/\max(f(S_j))$, $j = 1, 2, \ldots, n$; $f(S_i)$ is the fitness value of chromosome S_i; and $f_{\text{sum}} = \sum f'(S_j)$, $j = 1, 2, \ldots, n$. The procedure of selection is as follows:

Step 1: Compute the selection probability P_i for chromosome i.
Step 2: Compute cumulative probability $\text{CP}_r = \sum P_i, i = 1, 2, \ldots, r$.
Step 3: Let $k = 1$.
Step 4: Generate a random number, η, uniformly distributed from the range 0.0 to 1.0. If $\text{CP}_{r-1} < \eta \leq \text{CP}_i$, select chromosome i as one of the elements in the next generation.
Step 5: Set $k = k + 1$.
Step 6: If $k = N$ (size of population), terminate; otherwise, go to Step 4.

9.5.3.2 Crossover

Definition 9.7. (Connective Point). Given a collection of SubOs C_B and two triples $t_1 = \langle A, R, B \rangle$, $t_2 = \langle D, S, E \rangle$, we have $C_1 = E$ and $C_2 = A$ or $C_1 = B$ and $C_2 = D$. If C_1 and C_2 have one of the following semantic relationships:

- *Subsumption*: Concept C_1 is subsumed by a concept C_2 with respect to C_B if $C_1^I \subseteq C_2^I$ for every model I of C_B.
- *Equivalence*: Two concepts C_1 and C_2 are equivalent with respect to C_B if $C_1^I \subseteq C_2^I$ for every model I of C_B.
- *Disjointness*: Two concepts C_1 and C_2 are disjointed with respect to C_B if $C_1^I \cap C_2^I = \phi$ for every model I of C_B.

Then we say there is a connective point between t_1 and t_2.

A single-point, triple-based crossover operator is applied. The operator manipulates two chromosomes at the connective points where the triples in both chromosomes can be connected.

Assume the population size is N and the crossover rate is P_c. The crossover operation is as follows:

Step 1: Let $k = 1$.
Step 2: Randomly select two different chromosomes. Check the connective points between the two chromosomes and pick out all the concept pairs.
Step 3: Randomly select a connective point and divide each chromosome into two sets of triples at the point, one with a cache value no less than the cache value of the whole SubO, the other with a cache value less than the cache value of the

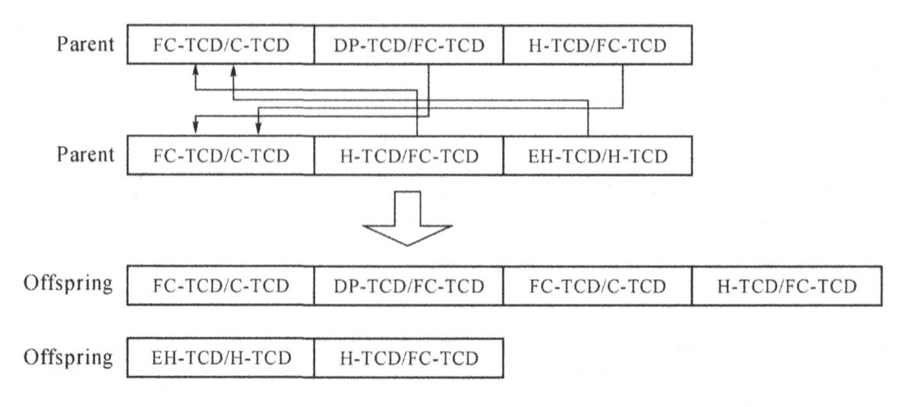

Figure 9.9 An example of the crossover operator.

whole SubO. Join the two parts with the higher value from the two chromosomes together and the other parts together.

Step 4: Set $k = k + 2$.

Step 5: If k is larger than $N \cdot P_c$, stop; otherwise, repeat Step 2.

For example, the two chromosomes in Figure 9.9 have four connective points. We can randomly select a connective point, for example, Functional Classification-TCD, divide the two parents into four parts, and then mate them to get two off-spring. The crossover operator can optimize the structure of parent SubOs and make it more compact. More importantly, the operator results in a better offspring and a worse offspring compared with the parents. The better offspring with high value will survive, while the other will die.

9.5.3.3 Mutation

A single-point, triple-based mutation operator is applied. The operator adds one chosen triple with high value from the ontology cache to the chromosome or removes a triple with low value from the chromosome.

Assume the population size is N and the mutation rate is P_m. The mutation operation is as follows:

Step 1: Let $k = 1$.

Step 2: Select a chromosome and generate a random integer with a uniform distribution of $(0, 1/P_m)$. If the number equals to 0, a mutation is conducted to the chromosome. Randomly select a triple from the set of triples with high cache value and append it to the chromosome or remove a triple with low cache value from the chromosome.

Step 3: Set $k = k + 1$.

Step 4: If k is larger than N, stop; otherwise repeat Step 2.

For example, to the chromosome in Figure 9.10, we can randomly select a triple from the triple list of the ontology cache and add it to the chromosome at a

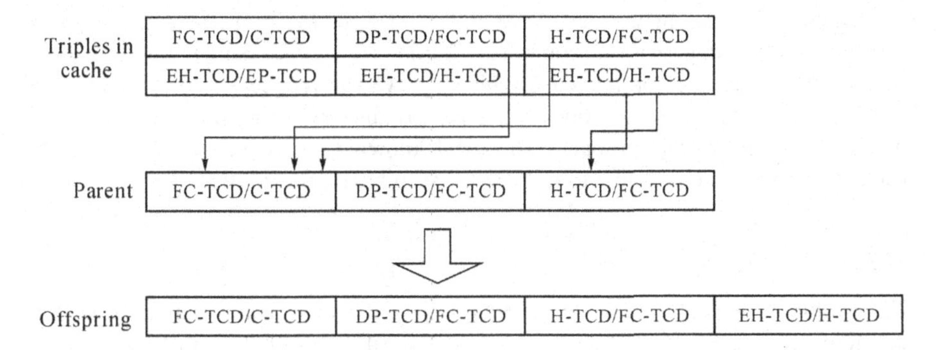

Figure 9.10 An example of the mutation operator.

connective point Hemostatics-TCD to get an offspring. The mutation operator enhances the chance for the evolution to jump out of suboptimization. Moreover, as the operator appends a triple to a chromosome with some probability, it also enhances the chance for the chromosome to be connected with other chromosomes.

9.5.4 Evolution Procedure

We use a GA to achieve SubO evolution based on the chromosome representation, fitness function, as well as genetic operators mentioned before. The evolution procedure is illustrated as follows:

Step 1: Warm up the ontology cache by extracting SubOs from the source ontology according to some questions.

Step 2: Start evolution if the volume of SubOs in the cache exceeds a predetermined level, i.e., about 90% of the cache space is occupied by SubOs.

Step 3: Encode SubOs in the cache as an initial population of chromosomes.

Step 4: After an initial population is generated, evolve the population based on a GA. The genetic process is performed iteratively until an optimal result is found.

Step 5: The GA carries out the genetic operators to generate offspring based on the initial population.

Step 6: Once a new generation is created, compare the chromosomes in the population and merge ones with high similarity.

Step 7: Evaluate the fitness value of the chromosomes in the population. Terminate if the overall fitness is higher than a threshold value; otherwise, go to Step 5.

Step 8: Decode the chromosomes in the result population to a set of SubOs and replace them with the original ones in the cache. Note that crossover and mutation can change the structure of a chromosome. When we decode a chromosome into a new SubO, the operation of extracting different parts from the original SubOs may be required.

The evolution algorithm has utilized semantics to optimize the ontology cache in several aspects. First, we use triple-based nonbinary encoding to represent SubOs as chromosomes, and the chromosome representation can preserve the semantics of SubOs. In the genetic operators of our GA, only those triples with semantic relationships can be combined. Before we replace SubOs out of cache, they have already been optimized. In this way, we can reduce loss of semantics in cache replacement. We can improve the performance of an ontology cache based on the SubO evolution approach. The local knowledge structure of the ontology cache becomes more adaptive to knowledge searching via evolution.

9.5.5 Consistency

The evolution of SubOs does not generate any new content compared to the source ontology. The result of evolution is a collection of new SubOs with a more compact and optimized structure. We can prove that the SubOs after evolution are still valid ones and thus consistent with the source ontology.

- *To the selection operator*: As the initial SubOs in an ontology cache are extracted from the source ontology, they are valid. What the selection operator does is only selecting some SubOs from a collection of valid SubOs; therefore, the SubOs after selection are valid and consistent with the source ontology.
- *To the crossover operator*: According to the definition of crossover, the operator only manipulates two chromosomes at the connective points where the triples in both chromosomes can be connected. The operator always results in two SubOs. It is possible for the crossover operator to generate an invalid SubO, which is disconnected in structure. According to Eq. (9.3), the disconnected SubO gets a zero value in fitness evaluation. The SubOs with zero value cannot pass the next round of selection, in other words, they will be discarded in selection. Therefore, the SubOs after crossover are valid and consistent.
- *To the mutation operator*: It is trivial to show that mutation also generates valid SubOs. The operator adds one randomly chosen triple with high value from the ontology cache to the chromosome or removes a triple with low values from the chromosome. The appended triple is also from the source ontology, so the resulting SubO is valid and consistent.

9.6 Experiment and Evaluation

In this section, we examine the performance of the evolution approach using the TCM ontology. As the TCM ontology is very large (more than 10,000 concepts and 80,000 individuals are defined in the current knowledge base), we only select a category for TCD from the whole ontology as the experimental dataset. There are about 1572 concrete concepts in the selected ontology subset.

9.6.1 Experiment Design

An important issue that impacts the performance of the cache is the cache replacement policy. Such policies always apply a value function to each of the cached items and choose the items with the lowest values as victims. Therefore, we want to compare the proposed evolution approach with the following traditional cache replacement policies: first in, first out (FIFO) [33] and least recently used (LRU) [34], as well as Semantic Caching.

With the FIFO policy, we record the order in which we put a SubO into the cache, and we always take the first one as the victim for replacement (see Figure 9.11A). With the LRU policy, we record the usage of each SubO in the cache. As long as a SubO is used to answer a question, we increase its usage. We always take the one least recently used as the victim (see Figure 9.11B).

We implement the Semantic Caching policy based on the method presented in Ref. [24]. As we mentioned before, SubO caching can be treated as a type of semantic caching. In semantic caching, a SubO is a semantic region for ontology triples. A question can be split into two disjoint portions: one that can be completely answered using the triples of SubOs present in the ontology cache (probe query), and another that requires triples to be extracted from the source ontology (remainder query). When a question intersects a SubO in the cache, the SubO gets split into two smaller disjointed semantic regions, one of which is the intersection of the SubO and the question, and the other is the difference in the SubO with respect to the question (see Figure 9.11C). Triples brought into the cache as the result of a remainder query also form a new semantic region. In this way, semantic regions will be merged according to their distance. Besides, the replacement policy of semantic regions or SubOs in semantic caching is based on LRU.

We have implemented four simple agents, each with an ontology cache and a cache replacement policy. The experimental setting and the major parameters are shown in Table 9.2. With each agent, we perform the following procedure:

Step 1: Warm up the ontology cache by extracting SubOs from the TCM ontology according to some questions.

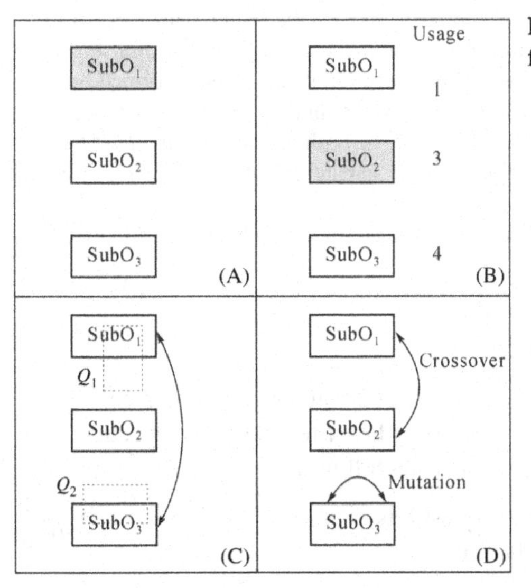

Figure 9.11 The four different policies for SubO cache.

Table 9.2 The Experimental Setting and Major Parameters

Parameter	Value
Size of the question set (N)	200
Size of question section (N_s)	20
Expired rate of the cache value (λ)	0.5
Cache space (N_c)	50
Depth of SubO extraction (d)	8
Evolution level (L_e)	0.9
Size of the population (N_p)	50
Crossover rate (P_c)	0.1
Mutation rate (P_m)	0.01

Step 2: Generate a set of questions based on the TCM ontology. Questions are generated automatically based on the TCM ontology. In order to simulate the manner users submit questions to a knowledge-based system, we generate questions according to the following rules:

- *Random generation*: There are many subsumption paths and relation paths in the TCM ontology; each can be used to make a question. Therefore, we can generate a collection of questions randomly based on the subsumption paths and the relation paths in the ontology.
- *Relevant generation*: Moreover, in order to simulate the locality of users' knowledge requests, we should not generate questions totally randomly. Instead, we generate one or more relevant questions after we have generated a question randomly.

Step 3: Submit the questions to the agent section by section. The agent answers questions by reusing existing SubOs in its ontology cache. Questions are selected randomly.

Step 4: Check the free space of the ontology cache. To the agent with the Evolution policy, perform evolution if the cache is nearly full. To the agent with the Semantic Caching policy, perform merging of regions if the cache is nearly full. For other ontology caches, just skip this step. The process of evolution is performed between the intervals of questions. This is to some extent similar with the behavior of pre-fetch, so we do not take into account the time of evolution when computing the cost of cache response. Merging of regions in the Semantic Caching policy is performed in the same way.

Step 5: For each agent, calculate the average response time and hit-ratio of its ontology cache every section.

9.6.2 Compare Cache Performance

The primary metrics used to evaluate the performance of an ontology cache are hit-ratio and response time. A cache with higher hit-ratio and lower response time is better.

The response time for the ontology cache is as follows:

$$T_{\text{response}} = \begin{cases} T_{\text{search}} + T_{\text{answer}} & \text{if a request hits} \\ T_{\text{penalty}} + T_{\text{answer}} & \text{otherwise} \end{cases} \tag{9.6}$$

where $T_{penalty} = T_{extraction} + T_{store}$, T_{search} is the time to search and retrieve a SubO from the ontology cache, T_{answer} is the time to answer a question by reusing a SubO, $T_{extraction}$ is the time to extract a SubO from the source ontology, and T_{store} is the time to store a SubO in the ontology cache.

As the TCM ontology is stored in a relational database as a Jena [35] model and the ontology cache is implemented in the memory, $T_{penalty} \gg T_{search}$ and $T_{penalty} \gg T_{answer}$. Therefore, we can ignore T_{search} and T_{answer} when computing the response time. The results presented here are a small but representative set of the experiments we have run.

The hit-ratio for the ontology cache is as follows:

$$H_{rate} = N_{reuse} / \Delta N \tag{9.7}$$

where N_{reuse} is the number of reused existing SubOs in the cache and ΔN is the section of questions answered within a time span.

Figure 9.12 shows the average hit-ratios for the four agents with four policies that provide a different performance across the range of questions. From the figure, we can see that Evolution does the best, followed by Semantic Caching, LRU, and then FIFO.

In the beginning, there is enough free space for SubOs in the Evolution cache, thus far from performing evolution. Therefore, the performance of the Evolution cache is fairly close to the others. After answering a batch of questions, the Evolution cache starts to evolve the SubOs. When an evolution occurs (at 60 questions), the performance of the ontology cache can be improved a lot. After several generations of evolution, the Evolution cache performs much better than the LRU cache and the FIFO cache.

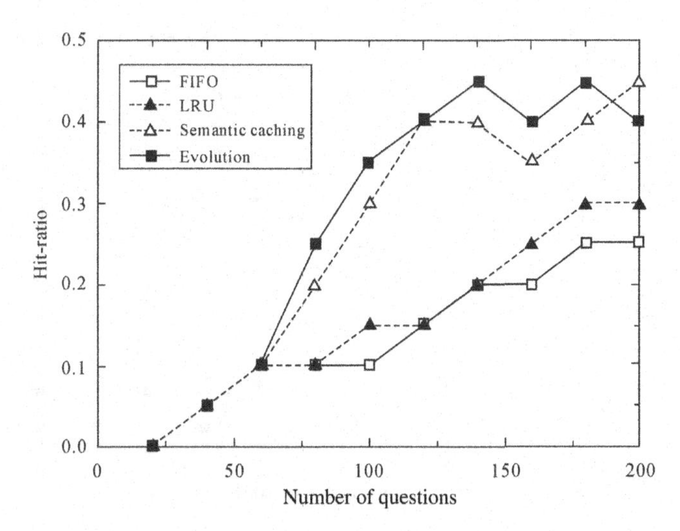

Figure 9.12 Hit-ratios for four cache policies.

Figure 9.13 shows the average response time of the four agents with different cache policies. As response time is directly related to hit-ratio, the result in Figure 9.13 is similar to that in Figure 9.12. The Evolution cache performs best in terms of response time. Generally speaking, the average response time for the Evolution cache is decreased by evolution.

According to this experiment, the Evolution policy performs better than others. There are semantic relations both within and among SubOs in the cache. The granularities of various algorithms used to manipulate SubOs are different. Both FIFO and LRU take the entire SubO as the cache object. This placement is performed in terms of SubOs. Therefore, both of them ignore the internal semantics of SubO. In contrast, both Semantic Caching and Evolution optimize the ontology cache in terms of triples. The advantage of FIFO and LRU policy is that they both are easy to be implemented, while Semantic Caching and Evolution are more complex. The major weakness of FIFO and LRU is that they do not make use of semantics within and among SubOs. Both Evolution and Semantic Caching optimize the structure of the ontology cache by making use of the semantics. Therefore, Evolution and Semantic Caching perform better than FIFO and LRU. Besides, the efficiency of GA also contributes to better performance of the Evolution policy. The Evolution mechanism of GA is more efficient for solving optimization problems like that of the ontology cache, compared to the splitting and merging of the Semantic Caching policy.

In this experiment, it can be seen that Semantic Caching and Evolution provide a similar performance across the range of question sections. Therefore, the semantic caching method can be considered as a candidate for the GA-based method in our SubO model. However, as the semantic caching method is proposed for database caching, we should extend it to the knowledge aspect in order to support ontology caching. A practical alternative is to combine the GA-based method with

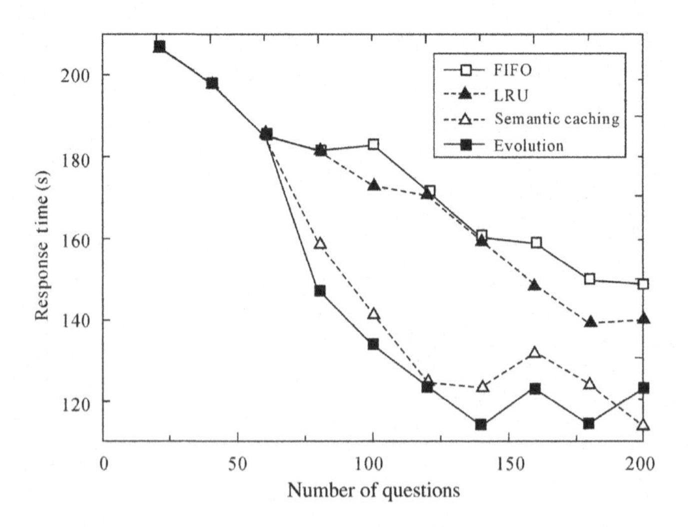

Figure 9.13 Response time for four cache policies.

semantic caching, as these two methods have their own focus and we can try to make use of the advantages of both.

9.6.3 Knowledge Structure

We also analyzed the context descriptions (B_c) of the evolved SubOs in the ontology cache to find out that the domain knowledge in the SubOs clusters in several contextual concepts. It means to see if the agent with an ontology cache becomes specialized in answering TCM-specific questions in some contexts as a result of evolution. The top-level concept of the selective part of the TCM ontology is TCD, which has seven subconcepts: Pharmacy Engineering-TCD (PE-TCD), Commoditylogy-TCD, Basic Theory-TCD (BT-TCD), Science of Processing-TCD, Pharmaceutics-TCD (Ps-TCD), Pharmacology-TCD, and Science of Identification-TCD (SI-TCD).

We can categorize the concepts in the ontology into seven groups, corresponding to the seven second-level concepts of the ontology. If a concept is the subconcept of a second-level concept in the ontology, then it is grouped in that second-level concept. Therefore, the concept sets of the evolved SubOs are also grouped in different concepts. Given a SubO $B = \langle B_c, B_t, B_a, O \rangle$, if a concept in B_c is the subconcept of a second-level concept in the ontology, it is grouped in that second-level concept.

Figure 9.14 depicts the group distribution of the concept sets for the evolved SubOs. The *y* axis of the figure shows the number of concepts in each group. Therefore, we can get the number of SubOs that have relevant knowledge for different concept sets. From the figure, it can be seen that the SubOs are mainly related to Ps-TCD, SI-TCD, BT-TCD, and PE-TCD. In other words, the agent is more specialized in answering questions in those contexts.

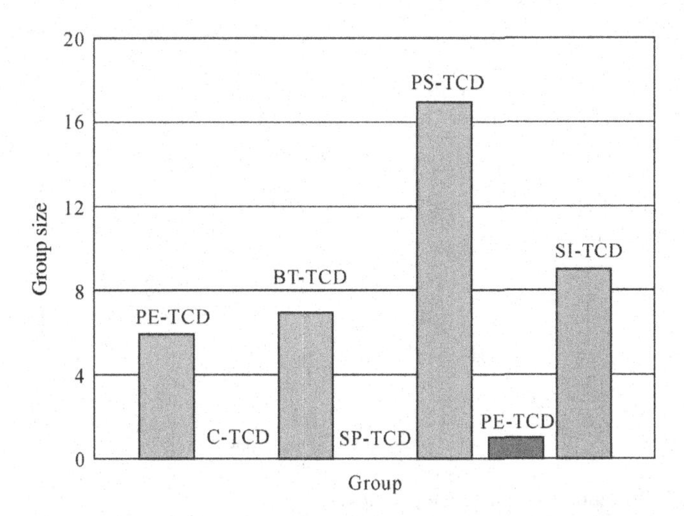

Figure 9.14 The group distribution of the context descriptions for the evolved SubOs.

9.6.4 Traversal Depth for SubO Extraction

In this experiment, we set an initial traversal depth of 8 and get the results above. We can change the depth to different values and then evaluate the performance of the Evolution cache with the same set of questions. Figure 9.15 shows the average hit-ratios of the Evolution cache with four different depth values for SubO extraction. From the figure, we can see that the traversal depth for extracting SubOs also impacts on the performance of SubO evolution. It can be seen that the traversal depth implies the SubO size and the ontology cache with greater depth performs better. A larger SubO will have a high chance to be reused by the space locality, and therefore the cache performance is improved.

When determining the depth for extracting the SubO, it is similar to the size of the cache block to some extent. The major difference is that a larger SubO is better not only for improving the hit-ratio by taking the advantage of space locality but also for improving the chance of crossover for chromosomes. An extreme case is that the depth of extraction for all SubOs is set to 1, and it is obvious that the chromosomes have few chances to cross over with each other. However, the depth should not be so deep that it causes the miss penalty and conflict misses or even capacity misses if the cache space itself is small. Besides, a great depth also increases the cost of extracting a SubO from the source ontology. When we increase the depth value from 8 to 10, the performance is not improved as much as we expected. The cache with a depth of 8 performs very closely to the one with a depth of 10, as can be seen from Figure 9.15.

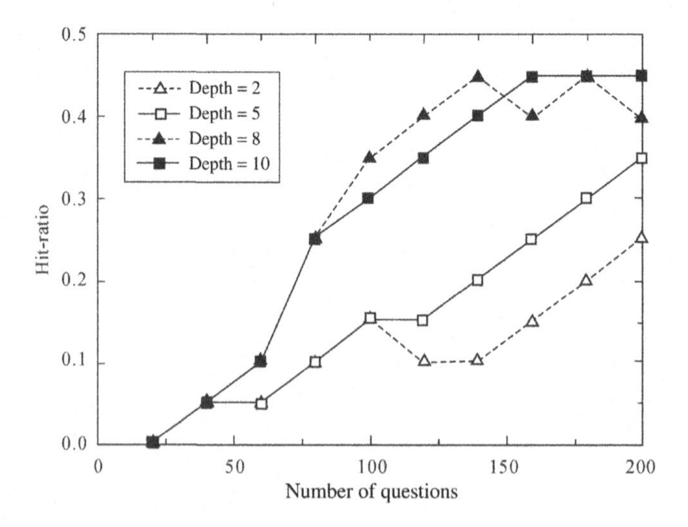

Figure 9.15 Hit-ratios for four cache policies.

9.7 Related Work

Compared with our work, the work related to our approach may be broken down into the following categories:

Ontology modularity and segmentation: The idea of extracting a subset of a larger ontology is referred to by many different names: views, segments, extracts, islands, modules, packages, partitions, and so on. This research topic is mainly about segmenting or extracting modules from large ontologies to satisfy application requirements.

Noy and Musen [36] argued that users need the ability to extract self-contained portions of ontologies to be used in their applications. They defined a portion of an ontology as an ontology view, which is similar to the view concept in databases for providing a self-contained portion of a resource. They proposed an approach for specifying ontology views through traversal of concepts and presented a strategy for maintaining traversal views through ontology evolution. Seidenberg and Rector [37] proposed some algorithms for extracting relevant segments from large DL ontologies. Their approach took advantage of the detailed semantics captured within an OWL ontology to produce highly relevant segments. They interpreted super-classes and quantified restriction as links between classes. The algorithm traverses the graph structure of an ontology to extract a segmentation. They also proposed to constrain the segment size by property filtering and depth limiting. They evaluated the approach with a medical ontology. Grau et al. [25] investigated the modularity of web ontologies and provided a notion of modularity grounded on the semantics of OWL-DL. They defined the notation of locally correct and complete to capture the feature of ontology modularity. They presented an algorithm for automatically identifying and extracting modules from OWL-DL ontologies. The main idea of the algorithm is to generate a partitioning of an input ontology, which is represented as a directed labeled graph (the partitioning graph) and use the graph to find the module for each term in the ontology. Doran [38] focused on ontology modularization for reuse. The aim of his research was to reduce the size of the ontology being imported to the part that is deemed relevant for the new application. He wanted to separate only part of an ontology, in a way such that a new ontology is generated, centered around a specific concept, which is correct and self-contained, and could be reused instead of the original one. A tool named ModTool has been implemented to provide an intuitive GUI for generating ontology modules. ModTool recursively applied a set of rules to generate an ontology module.

Ontology context: This research topic is mainly about the contextual phenomena of ontologies or DL knowledge bases. According to Ref. [16], contexts are local (where local is intended here to imply not shared) models that encode a party's view of a domain. Bouquet et al. [39] showed how ontologies could be contextualized, thus acquiring certain useful properties that a purely shared approach cannot provide. An ontology is contextualized or it is a contextual ontology when its contents are kept local, and therefore not shared with other ontologies and mapped with the contents of other ontologies via explicit (context) mappings. They proposed Context OWL (C-OWL), a language whose syntax and semantics had been obtained by extending the OWL syntax and semantics to allow for the representation of contextual ontologies. Zhao et al. [40] proposed a new framework for managing autonomy in a set of cooperating ontologies to overcome the shortcoming of C-OWL. In their framework, each ontology entity may have its meaning assigned either locally or globally, thus enabling semantic cooperation between multiple ontologies. Each ontology has a subjective semantics based on local interpretation and a foreign semantics based on semantic binding to neighboring ontologies. Mena et al. [23] expected multiple ontologies to be used as different world views and proposed to use multiple ontologies as the paradigm for information access on the web. They also presented an approach for estimating loss of information based on navigation of ontological terms.

Ontology or model algebra: This research topic concerns the formal algebra for ontology or complex knowledge models. Wiederhold [41] presented an algebra for building composed ontologies from domains that have ontological differences. Given a formal Domain Knowledge Base model containing matching rules that define sharable terms, he defined the algebra that contained a collection of binary operations (intersection, union, and difference) among domains. The algebra can provide a basis for interrogating multiple databases which are semantically disjointed but where a shared knowledge base has been established. Bernstein et al. [42] proposed an approach for complex model management. They tried to make database systems easier to use for these applications by making model and model mapping first-class objects with special operations that simplify their use. Their major contribution is a sketch of a proposed data model. The data model consists of formal, object-oriented structures for representing models and model mappings, and of high-level algebraic operations like matching, differencing, merging, selection, inversion, and instantiation.

Compared with existing research efforts, our approach mainly focuses on reusing and evolving context-specific portions from large-scale ontologies to improve the overall performance of ontology-based systems. Moreover, we present the subontology evolution approach based on a GA for ontology reuse. In contrast, we just proposed the concepts of SubO and SubO evolution, but no concrete algorithm was implemented in our previously published work [20].

9.8 Conclusions

In this study, we introduced a large-scale web ontology for TCM. We argued that to meet the on-demand and scalability requirement, ontology-based systems should go beyond using ontology statically and be able to self-evolve for their local domain knowledge. We presented a dynamic evolution approach for reusing large-scale ontology. In particular, we proposed the concept of subontology to represent context-specific portions from the TCM ontology and reuse them in a cache-like repository to support knowledge searching for reasoning. We showed that the main factor that impacts the performance of the ontology cache is the knowledge structure of SubOs. We proposed to evolve SubOs in the ontology cache to optimize the knowledge structure. We explained the details about the SubO evolution approach based on a GA. Triple-based encoding, fitness function, and genetic operators were used to support the evolution. We have conducted an experiment to evaluate the proposed evolution approach with the TCM ontology and obtained promising results.

There is still much room for improvement of the proposed SubO evolution approach. Future research issues include (1) improving the evolution algorithm by using more semantic information in the ontology, especially if we can improve comparing SubOs by using semantic information loss, (2) extending the approach to a multi-agent environment and studying the knowledge distribution in such an environment, and (3) evaluating the evolution algorithm with semantic caching and making use of the approach to improving the SubO model.

References

[1] Cai JF. An historical overview of traditional Chinese medicine and ancient Chinese medical ethics. Ethik in der Medizin 1998;10:84–91.

[2] Mao YX, Wu ZH, Xu Z, et al. Interactive semantic-based visualization environment for traditional chinese medicine information. In: Proceedings of seventh Asia–Pacific web conference; 2005. p. 950–59.

[3] Gruber T. A translation approach to portable ontology specifications. Knowl Acquis 1993;5(2):199–220.

[4] Heijst GV, Schreiber AT, Wielinga BJ. Using explicit ontologies in KBS development. Int J Hum Comput Stud 1997;46(2/3):183–292.

[5] Berners-Lee T, Hendler J, Lassila O. The semantic web. Scientific American 2001;284(5):34–43.

[6] Bodenreider O. Unified medical language system (UMLS): integrating biomedical terminology. Nucl Acids Res 2004;32(D):267–70.

[7] Ashburner M, Ball CA, Blake JA, et al. Gene ontology: tool for the unification of biology. The gene ontology consortium. Nat Genet 2000;25:25–9.

[8] Whetzel PL, Parkinson H, Causton HC, Patricia L, et al. The MGED ontology: a resource for semantics-based description of microarray experiments. Bioinformatics 2006;22(7):866–73.

[9] Yu AC. A method in biomedical ontology. J Biomed Inform 2006;39(2006):252–66.

[10] Fahmi I, Zhang J, Ellermann H, et al. SWHi system description: a case study in information retrieval, inference, and visualization in the semantic web. In: Proceedings of ESWC; 2007. p. 769–78.

[11] Lucene.<http://lucene.apache.org/>; 2010.

[12] TCMLS.<http://sharelab.cintcm.com/TCMLSOntoEdit/demo/test/brow-ser.html/>; 2010.

[13] Klyne G, Carroll JJ. Resource description framework (RDF): concepts and abstract syntax. W3C recommendation, world wide web consortium (W3C); 2004. Available from: <http://www.w3.org/TR/2004/REC-rdf-concepts-20040210/>; 2010.

[14] McGuinness DL, Harmelen FV. OWL Web ontology language overview. W3C recommendation, world wide web consortium (W3C); 2004. Available from: <http://www.w3.org/TR/2004/REC-owl-features-20040210/>.

[15] Qooxdoo. <http://www.qooxdoo.org/>; 2010.
[16] Ghidini C, Giunchiglia F. Local models semantics, or contextual reasoning = locality + compatibility. Artif Intell 2001;127:221–59.
[17] Lenat DB. Cyc: a large-scale investment in knowledge infrastructure. Commun ACM 1995;38(11):33–38.
[18] Blair P, Guha RV, Pratt W. Microtheories: an ontological engineer's guide, MCC Technical Report, CYC-050-92, March 1992.
[19] Chen HJ, Wang YM, Wang H, et al. Towards a semantic web of relational databases: a practical semantic toolkit and an in-use case from traditional chinese medicine. In: Proceedings of international semantic web conference; 2006. p. 750–63.
[20] Mao YX, Cheung WK, Wu ZH, et al. Dynamic sub-ontology evolution for collaborative problem-solving. In: Proceedings of AAAI fall symposium v FS-05-01; 2005. p. 1–8.
[21] Baader AF, Nutt W. Basic description logics. In: Baader F, Calvanese D, McGuinness D, et al., editors. The description logic handbook: theory, implementation, and applications. Cambridge University Press; 2003.
[22] Levenshtein IV. Binary codes capable of correcting deletions, insertions, and reversals. Cybern Control Theory 1966;10(8):707–10.
[23] Mena E, Kashyap V, Illaramendi A, et al. Imprecise answers in distributed environments: estimation of information loss for multiple ontology based query processing. Int J Coop Inf Syst 2000;9(4):50–56 [Special issue on Intelligent Integration of Information].
[24] Dar S, Franklin M, Jonsson B, et al. Semantic data caching and replacement. In: Proceedings of international conference on very large data bases; 1996. p. 330–41.
[25] Grau BC, Parsia B, Sirin E, et al. Modularity and web ontologies. In: Proceedings of international conference on the principles of knowledge representation and reasoning; 2006. p. 198–209.
[26] Baader F, Sattler U. An overview of tableau algorithms for description logics. Studia Logica 2001;69:5–40.
[27] Schmidt-Schauß M, Smolka G. Attributive concept descriptions with complements. Artif Intell 1991;48(1):1–26.
[28] Klein M, Fensel D. Ontology versioning for the semantic web. In: Proceedings of international semantic web working symposium; 2001. p. 483–93.
[29] Noy NF, Klein M. Ontology evolution: not the same as schema evolution. SMI technical report SMI-2002-0926, Stanford Medical Informatics; 2002.
[30] Holland JH. Adaption in natural and artificial systems. Cambridge: MIT Press; 1975.
[31] Chamber L. Practical handbook of genetic algorithms. Florida: CRC Press; 1995.
[32] Goldberg DE. Genetic algorithms in search, optimization, and machine learning. Boston: Addison-Wesley; 1989.
[33] Smith AJ. Cache memories. Comput Surv 1982;14(3).
[34] Edward G, Coffman J, Peter JD. Operating systems theory. New Jersey: Prentice-Hall; 1973.
[35] Jena. <http://jena.sourceforge.net/>; 2010.
[36] Noy NF, Musen MA. Specifying ontology views by traversal. In: Proceedings of third international semantic web conference; 2004. p. 713–25.
[37] Seidenberg J, Rector A. Web ontology segmentation: analysis, classification and use. In: Proceedings of international conference on world wide web; 2006. p. 13–22.
[38] Doran P. Ontology reuse via ontology modularisation. In: Proceedings of knowledge web, Ph.D. symposium. Budva, Montenegro; 2006.

[39] Bouquet P, Giunchiglia F, Harmelen FV, et al. C-OWL: contextualizing ontologies. In: Proceedings of international SEMANTIC web conference; 2003. p. 164–79.
[40] Zhao Y, Wang K, Topor R, et al. Semantic cooperation and knowledge reuse by using autonomous ontology. Technical report DIT-07026, Informatica etelecomu-nicazioni, University of Trento; 2007. Available from: <http://semanticweb.org/wiki/Semantic-cooperation-and-knowledge-reuse-by-using-autonomous-ontologies>.
[41] Wiederhold G. An algebra for ontology composition. ARPA and Stanford University; 1994. Available from: <http://infolabstanford.edu/pub/gio/1994/onto.ps>.
[42] Bernstein PA, Levy AY, Pottinger RA. A vision for management of complex models. SIGMOD Rec 2000;29(4):55–63.

10 Semantic Association Mining for Traditional Chinese Medicine

10.1 Introduction

Domain-driven data mining (D^3M) is an emerging paradigm of data mining (DM) [1] that "meta-synthesizes" various intelligence resources in order to support actionable knowledge discovery in complex domain problems [2–4]. Integrated Medicine (IM) is a suitable application domain for D^3M. As an emerging interdisciplinary subject, IM is intended to synthesize the complementary knowledge elements from modern biomedicine and various systems of Traditional Medicine [5] for maximizing therapeutic benefits to patients [6,7]. As a characteristic activity of IM, collaboration in knowledge discovery is beneficial yet challenging, largely due to barriers to interdisciplinary communication. The D^3M paradigm can facilitate IM studies by (1) formalizing and synthesizing complementary knowledge elements from diverse domains using ontologies and (2) facilitating the collaborative reasoning among a network of human users and intelligent agents. A series of seminal works has made progress in ontology engineering [8], heterogeneous data integration [9], and knowledge representation [10], management [11], and discovery [12,13] for the IM domain, yet these resulting intelligence resources need to be "meta-synthesized" to build domain-driven intelligent systems satisfying the needs of IM domain experts.

10.1.1 The Semantic Web for Collaborative Knowledge Discovery

The Semantic Web [14] could facilitate collaborative D^3M through the semantic integration of intelligence resources from different communities driven by domain-specific problems. In particular, it is successfully applied to biomedicine, to connect a plurality of knowledge and data elements and support their translation across domains [14,15]. These applications commonly use biomedical ontologies [16] to integrate distributed and heterogeneous databases, for effective retrieval and serendipitous reuse [17–20]. In addition, intelligent agents with reasoning capabilities could be applied to the Semantic Web of data in order to solve complex application problems and discover novel knowledge [21]. To realize this potential, however, feasible collaboration mechanisms among multiple agents need to be further investigated.

As our major concern, the Semantic Web is especially suitable for domain-driven relationship mining. Semantic Association Mining (SAM) is proposed to infer implicit or latent relationships between arbitrary resources based on patterns

Modern Computational Approaches To Traditional Chinese Medicine. DOI: http://dx.doi.org/10.1016/B978-0-12-398510-1.00010-8

discovered from Semantic Web data [20,22,23]. SAM processes are typically based on ontologies (as the scheme for domain knowledge representation and semantic integration of resources) and driven by specific application problems, and therefore can be categorized as the D^{3}M. Here, we define Semantic Association (SA) as the representation of rich knowledge about binary relationships using the Semantic Web data model [22]. In the Resource Description Framework (RDF) [24], an SA is represented as a statement in the form of (*Subject*, *Property*, *Object*) triple, where the Property defines the type of the link between the Subject and the Object. The RDF Graph, as a set of statements, is a graph-theoretic model suitable for the representation and merging of SAs. The patterns discovered by SAM are typically represented as RDF subgraphs, and SAM can be implemented as *subgraph extracting queries* against RDF Graph models [23]. SAM techniques seem to have huge potential for relationship mining involving interdisciplinary collaboration [21].

10.1.2 The Motivating Story

We present a case of collaborative SAM in traditional Chinese medicine (TCM) [5], in which in-use herbal therapies are analyzed to inspire modern drug discovery [25]. Here, an analyst discovers the frequent use of a TCM herb named *Huperzia serrata* (HS) to treat the TCM Syndromes [13] that are associated with the symptom of Memory Loss, which leads to the hypothesis that "HS seems to affect Memory Loss." However, scientific validation for the hypothesis goes beyond the scope of TCM, so that the hypothesis is published as an SA to solicit evidence. Scientists from disciplines such as pharmaceutics, biomedicine, or chemistry can then publish evidence to prove/falsify this hypothesis. Figure 10.1 shows a thread

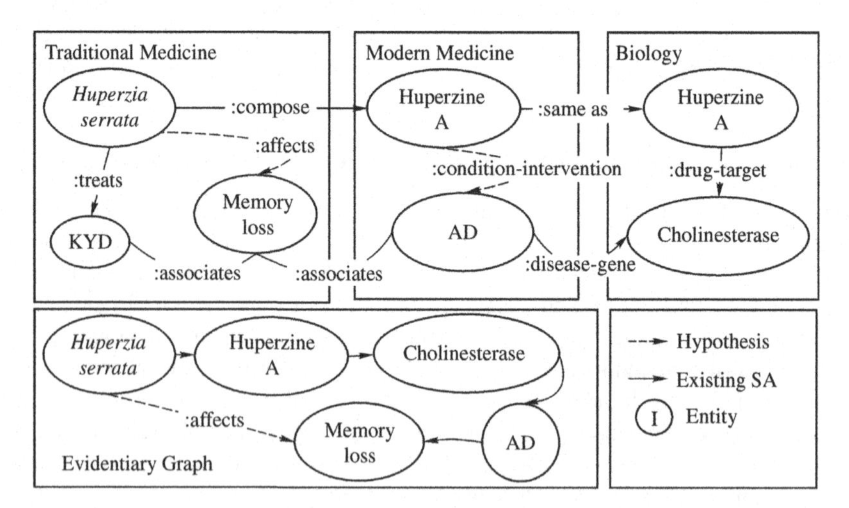

Figure 10.1 SAs from three domains (Traditional Medicine, Modern Medicine, and Biology) are synthesized into an Evidentiary Graph through the publication and validation of hypotheses.

of evidence: (1) HS contains the bioactive chemical Huperzine A [26]; (2) Huperzine A is an inhibitor of the enzyme cholinesterase, which is a drug target for Alzheimer's disease (AD) [27,28]; (3) AD is often associated with Memory Loss [29]. To support this type of analysis, complementary knowledge elements (hypotheses and evidence) should be extracted from distributed sources and published so that their relationships can be established and reasoned.

10.1.3 HerbNet: The Knowledge Network for Herbal Medicine

To support the above application, we intend to meta-synthesize a knowledge network—the HerbNet—from intelligent resources such as ontologies, data sources, as well as reasoning and learning capabilities. As shown in Figure 10.2, the HerbNet is twofold: (1) the "Graph of Things" [30] encapsulates the (hypothetical) relationships under investigation, with nodes representing medical entities and edges representing SAs annotated with medical evidence; (2) the "Web of Documents" refers to the network of web-based intelligent resources containing the evidence for SAs. The application requires that the "Graph of Things" be synchronized with the "Web of Documents," navigable and editable by multiple parties, and also capable of answering domain-specific complex problems. For such a knowledge network to emerge, SAs should be derived from intelligent resources and aggregated into a graph of things while retaining their contexts and connecting to their evidence for justification and validation.

To fuse the HerbNet, we propose a hypothesis-driven framework for a multitude of agents to collaborate in discovering latent SAs on the Semantic Web. We define a SA as a hypothesis, which is essentially a statement whose truthfulness depends on the existence of certain relevant evidence. On the other hand, an evidence is open or uncertain if its own reliability depends on the validity of some open hypotheses. The mutual dependency between hypotheses and (open) evidence gives rise to an evidentiary citation network. An agent can publish a predicted yet unproved SA as a hypothesis H, send H to a selective set of neighbors who might

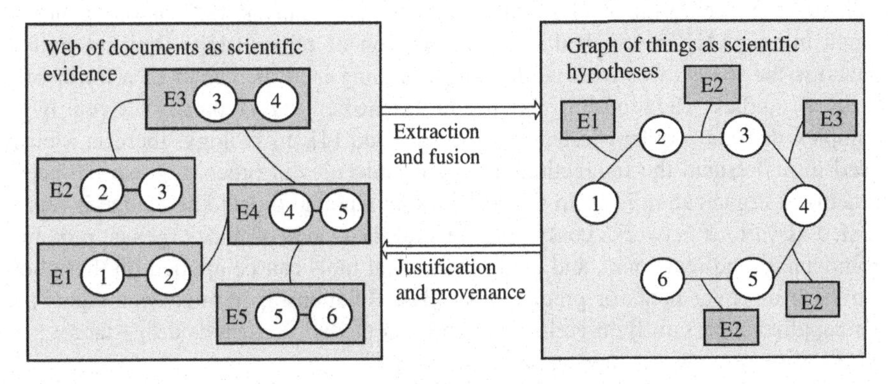

Figure 10.2 Implementation of the HerbNet as an ontology-based hypothesis-evidence dual network, instead of a static dataset.

be interested in it and devoted to solve it, and then periodically search for the evidence of H that other agents have published. Each agent, upon solving a hypothesis, can make his own decision on what (partial) evidence to provide and what (derivative) hypotheses to propose, in order to influence the direction of mining and contribute to the final solution to the hypothesis. We use ontology-based views to express the hypothesis-answering capacity of agents, so that agents can dispatch the right hypotheses to related agents to solve. This framework allows a knowledge network to emerge through the communication of SAs by a multitude of agents in terms of hypothesis and evidence.

10.1.4 Paper Organization

The rest of the chapter is organized as follows: Section 10.2 introduces the related work. Section 10.3 formulates the methodology of hypothesis-driven SAM, and Section 10.4 reports the simulation results regarding the effectiveness, scalability, and performance of our methodology. Section 10.5 reports our empirical experiments in actual cases of integrative medicine, such as the interpretation of TCM formulae and analysis of the herb–drug interaction network. Concluding remarks are given in the final section.

10.2 Related Work

As an emerging paradigm of DM, the D^3M possesses characteristics such as Problem-oriented, Ontology-based, Intelligent Data, Actionable Knowledge, and Intelligence Meta-synthesis [2–4]. In this section, we inspect the advantages and limitations of related work using the criteria of D^3M.

10.2.1 Domain-Driven Relationship Mining for Biomedicine

Relationship mining (also called link mining or link analysis [23]) is one genre of graph mining [31] in which the characterization of relationships (but not individuals) is the focus of concern. Relationship mining is at the center of a lot of biomedical studies. Relationship mining in biomedicine is typically driven by a complex domain problem instead of specific data [4]. In biology, there is a clear need to understand the interactions between molecules in order to grasp the cell's functional organization [32]. In network biology, the biological knowledge is represented as various networks comprising several thousands of nodes (genes, proteins, compounds, and reactions), and graph-theoretical tools can be applied on these networks to discover paths or predict links [32]. Relationship mining techniques are also applied successfully in various networks in medicine, such as drug–target networks, disease networks, and social networks [33].

Some work has been performed on relationship mining in Integrative Medicine (IM) [5,34]. In IM, physicians examine the patient as an integrated system and make holistic therapeutic strategies to promote health and enhance the quality of life, rather than focusing on specific diseases or symptoms [5]. Here, the human body is viewed as a complex network of subsystems that interact or communicate with each other and with the environment [7]. Therefore, studying the complex relationships between various biomedical entities is a critical problem. In Ref. [13], the relationships between TCM Syndromes (the TCM notion of the morbid state of human body, simply "syndrome" from now on) and diseases are automatically learned from the literature. These relationships are then combined with the disease–gene relationships to generate (indirect) syndrome–gene relationships, which are transformed into syndrome-based gene networks that embody the functional knowledge of genes from the syndrome perspective. Also, the survey [12] introduces a thread of work that learns association rules between herbs from a set of formulae. The work, although in its infancy stage [12], demonstrates the usefulness of relationship mining methods in distilling valuable insights from a combined data repository of TCM and modern medicine.

In addition to discovering relationships from raw data (generated by high-throughput data-collection techniques [32] or clinical trials and cases), relationships can also be learned from textual documents [35]. In relationship extraction, agents try to understand the expression of relationships made directly by the authors and represent the relationships in a machine-understandable format [20,36]. In addition, some work tries to infer hypothetical relationships by the presence of other more explicit relationships [13,35].

Scientific exploration of relationships is often based on graph-theoretical models and uses graph mining [31] methods. Relationship discovery applications generally require a comprehensive view of knowledge and data distributed among people and organizations across geographical areas. In these studies, however, graphs are typically stored in databases, and graph mining methods are implemented on top of databases. The major drawback of databases is that the data and resulting knowledge cannot be published and shared in a standardized mechanism, therefore bringing obstacles to intelligence meta-synthesis across organizational boundaries. The Semantic Web, by providing a unified knowledge representation scheme and a data integration solution, can help improve the quality of relationship mining [37].

10.2.2 Linked Data on the Semantic Web

The Semantic Web can foster interdisciplinary communication by enabling data to be shared in a machine-understandable format across domain boundaries [14]. The effort of publishing and interlinking structured data on the Semantic Web will result into a global network of databases or the Linked Data [30,38]. The Linked Data extends the use of the Uniform Resource Identifier (URI) [39] to the identification of any significant objects, including concrete things, abstract entities, and generalized concepts (relation types and classes). An RDF document can be abstracted as a "Graph of Things," with each node corresponding to a URI or Literal, and an edge corresponding to an RDF statement. For example, the document {(: Hup. A, : inhibits, : Chol.), (: Chol., : relates, : Alzheimer's), (: Hup. A, : moderates, : Alzheimer's)} can be represented as the graph:

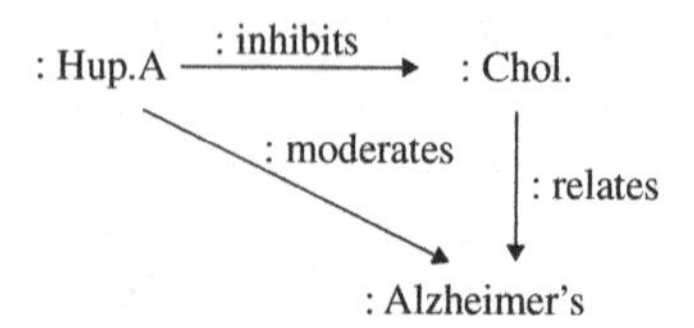

The Linked Data browser allows human users and software agents to look up the RDF document about a resource (using HTTP protocol) and to navigate and edit these "graphs of things" [30]. In addition, the Semantic Web can facilitate the merging of graphs into a large "multigraph." A graph is ground if it has no blank nodes [40]. The merger of two ground graphs G_1 and G_2, denoted as $G_1 + G_2$, is the set theoretical union of their sets of statements [40]. Within a "multigraph," a member graph can obtain a URI as its name for its provenance to be traced [41]. Therefore, Linked Data gives an elegant solution to connect data from different sources and domains.

The most promising application area of the Semantic Web of data is biomedicine, where the interoperability and integration of a plurality of knowledge and data are in great need [14,15]. One closely related application is the SWAN for the AD research community [18,19]. The SWAN also uses an ontology to formally represent the relationships between scientific hypotheses and their supporting evidence, as a basis for organizing the "knowledge ecosystem." However, whereas the SWAN Ontology [17] is a very elaborate OWL (Ontology Web Language) Ontology for scientific discourse in general, we focus on the problem of SAM and only introduce the necessary concepts (such as Hypothesis, Evidence, supports, and depends) into the existing RDF vocabulary in order to model the Evidentiary Citation Network.

In summary, the Semantic Web can provide a suitable infrastructure for D^{3}M, through (1) domain knowledge formalization using ontologies; (2) ontology-based integration of intelligence resources; and (3) making the data self-explanatory through the annotation of domain-specific features.

10.2.3 Semantic Association Mining

SAM possesses major characteristics of the D^{3}M such as ontology based, domain driven, and application specific. The major innovation of the SAM is to represent relationships in terms of a shared domain ontology, enabling any agents to reason about the relationships [20,23]. Biomedical ontologies [16], such as the MeSH and the Gene Ontology, can improve the quality of SAM through the following means:

1. Formally representing domain knowledge, data semantics, the problem-solving context, and relationships which have appeared in scientific publications [37];
2. Facilitating the (semi)-automated extraction of SAs from unstructured or semi-structured data of the specific domain [20,42];

3. Facilitating the integration of heterogeneous databases in order to map a more comprehensive network for relationship discovery [43].

By using ontologies, SAM can improve the accuracy, application relevance, and actionability of relationship discovery. However, SAM currently has two major limitations compared with the criteria of D^3M.

The major limitation of recent work is that SAM functions are typically conducted by one agent and fail to "meta-synthesize" the intelligence resources of multiple agents. In the work, SAM functions are implemented as queries against a centralized Semantic Web data model [20,22,23]. For example, the authors of Ref. [43] first download the data from distributive sources, and then populate the data into an RDF model to analyze the relationships within. Whereas the Semantic Web is developing into a diverse environment (involving ontologies and data distributed among multiple agents), designing a scalable infrastructure for supporting SAM in distributed environments is still an identified open problem [20,23]. Therefore, our major concern is to propose decentralized SAM algorithms that are scalable to a lot of distributed sources.

Second, although related work of SAM is typically presented in association with specific application problems, the actionability of SAM is still under evaluation. SAM seems to be potentially useful in social network analysis [43], national security [23], patent retrieval [20], and biomedicine [21]. However, the recent work lacks studies that enable domain experts to validate the effectiveness of proposed techniques in real-world applications [21,22]. Therefore, we emphasize testing the SAM approach on real-world medical data and evaluate the effectiveness of new techniques and the usefulness of discovered knowledge based on user studies.

10.3 Methods

In this section, we investigate how a multitude of agents can discover, validate, and rank complex relationships in a collaborative manner. We propose a hypothesis-driven SAM framework, in which agents extract patterns from their own semantic graphs driven by the need to solve published hypotheses. Agents can publish discovered patterns as partial evidence that depends on some other hypotheses. We will formulate the components of this methodology, such as hypothetical and evidentiary graphs, the semantic schema, and the generic procedures for the discovery and ranking of evidence.

10.3.1 Semantic Graph Model

The Semantic Graph (SG) Model is a graph-theoretical data model for abstracting the organization of knowledge bases in RDF and RDF Schema [20,44]. Assume there are two disjoint infinite sets U (URI references) and L (RDF Literals). A resource r is any entity that is identified by $u \in U$. A class c is a resource that represents a set of resources. A property p is a resource that represents a binary relation between class c_1 and class c_2, with c_1, c_2 as the domain and range of p. For example, the property "prescribes" may be defined between Physician (as the domain) and Drug (as the range). A link of type P can be represented as an RDF statement as follows:

Definition 10.1. (RDF Statement, Semantic Association, and Semantic Annotation). An RDF Statement is a triple $(s, p, o) \in U \times P \times (U \cup L)$ where s is the subject, p is the predicate, and o is the object. A Semantic Association (SA for short) is a statement $(s, p, o) \in U \times P \times U$ which asserts the existence of a relationship of type p between s and o. A Semantic Annotation for the resource r identified by s is a statement $(s, p, o) \in U \times P \times (U \cup L)$ which annotates r with value o of the type as specified by p. An annotation of numerical value is called a weight, and assigning a weight of type w and value v to the resource r identified by s is achieved by inserting a statement (s, w, v).

An RDF Graph is a directed labeled graph that represents a set of statements, in which each statement (s, p, o) is denoted by $s \xrightarrow{p} o$. In addition, a Named Graph is an RDF Graph assigned with a name that is specified by a URI reference [41], which is a feature of crucial use in the Evidentiary Citation Network. We discuss Named Graphs and refer to them as graphs by default from now on.

Definition 10.2. (Named RDF Graph). For a set D of statements, the Named RDF Graph is a pair (u, G) where u is a URI reference and G is a directed, labeled graph (V, E). In G, V is a set of vertices and E is a set of edges, such that for each statement (s, p, o) in D, s and o belongs to V, and (s, p, o) in E.

The SG model finally abstracts the Knowledge Space KS as a triple (U, G, A) where U is an infinite set of URIs, G is a set of graphs, and A is a set of agents that identify things with URIs and exchange knowledge elements as graphs. U has two subsets U_a and U_g that contain URIs for members of A and G, respectively.

10.3.2 Hypothesis and Hypothetical Graph

We define an SA problem as a published Hypothesis whose truthfulness is under investigation. A hypothesis is represented as a reified statement that is identified by a URI reference and annotated with numeric or literal values that deliver information about the hypothesis.

Definition 10.3. (Hypothesis). A hypothesis h is a quintuple (u, s, p, o, a), in which u is the URI reference for h; s, p, and o are h's subject, predicate, and object, respectively; and a is a set of annotations to h. Using the vocabulary including the class Hypothesis [hyp] as the sub-Class of rdf:Statement[stat], the properties rdf:subject[subj], rdf:predicate[pred], and rdf:object[obj] together with annotation properties such as dc:Creator[auth], we can represent h graphically as follows:

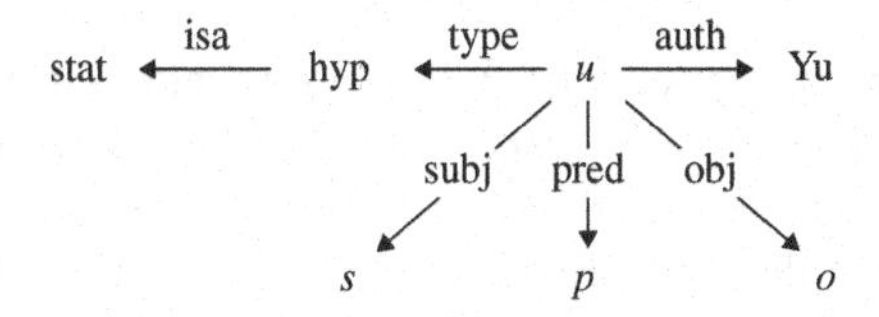

A hypothesis $h = (u, s, p, o, a)$ means that the problem that is identified by u looks at whether (s, o) belongs to the binary relation specified by p. A hypothesis can be uniquely identified by its URI reference or by its problem as is specified by a subject−predicate−object triple. Therefore, we can denote the above hypothesis $h(u)$ as $?(s, p, o)$ or $s \xrightarrow{?p} o$.

A meaningful hypothesis shall have necessary annotations such as: (1) the author; (2) human-friendly labels; (3) contextual information; (4) the evidence for proving or falsifying the hypothesis (to be discussed in the next section); (5) the weights for hypothesis ranking. A hypothesis can be expressed as a graph, which can be merged with other hypotheses and published as a named graph. We represent a named graph containing a set of hypotheses as a Hypothetical Graph:

Definition 10.4. (Hypothetical Graph). For a set D of hypotheses, a Hypothetical Graph hg is a direct graph (V, E) where V is a set of vertices and E is a set of edges, such that for each hypothesis $?(s, p, o)$ in D, we have $s, o \in V$ and $s \xrightarrow{?p} o$ in E.

10.3.3 Evidence and Evidentiary Graph

In general, an evidence can be any resource (documents, images, videos, or even persons) which can facilitate agents to achieve a belief (or denial) of a hypothesis. For our purpose, we focus on explicit and formal evidence that is expressed in Semantic Web languages. Such evidence is in itself a pattern such as a subgraph extracted from an agent's knowledge base, and evidence discovery can be seen as a subgraph extraction problem. In addition, the evidence can be open, meaning it can depend on some hypotheses.

Definition 10.5. (Evidence). An Evidence e is a triple (u, p, a), in which u is e's URI reference, p is a graph that specifies e's pattern, and a is a set of annotations to e. a may contain two hypothetical graphs: (1) sh that contains the hypotheses that e is intended to support; (2) dh that contains the hypotheses that e depends on (if e is uncertain).

In addition to the publication of hypotheses and evidence, agents also need to make assertions about their mutual dependency: (1) if an evidence e is proposed to support a hypothesis h, then the statement $(h, \text{depends}, e)$ can be asserted; (2) if a hypothesis h' is a member of the hypothetical graph dh of an evidence e', then the statement (e', depends, h') can be asserted. We propose an Evidentiary Graph to wrap up these dependency relationships.

Definition 10.6. (Evidentiary Graph). For two mutual exclusive sets H and E that contain hypotheses and evidence, respectively, an Evidentiary Graph is a directed bipartite graph (H, E, D) where H and E are two sets of vertices, and D is a set of edges corresponding to the mutual dependency relationships among members of H and members of E, so that for h in H, e in E:

$$(h, \text{depends}, e) \Leftrightarrow h \xrightarrow{\text{dep}} e \in D$$
$$(e, \text{depends}, h) \Leftrightarrow e \xrightarrow{\text{dep}} h \in D$$

The Evidentiary Graph enables the navigation of interrelated knowledge elements in their logical sequence. In order to enable navigation in the reverse direction, we introduce the property "supports" as the inverse property of "depends." We also define an Inverse Evidentiary Graph that is formed by reversing the direction of every edge in the corresponding Evidentiary Graph to wrap up the support relationships.

The Evidentiary Graph provides the contextual information for the evidence of a pattern. We define the set of neighboring hypotheses of evidence as the Semantic Annotation for the evidence of a pattern.

In this formulation, a Knowledge Network is represented by two graphs: a hypothetical graph and an evidentiary graph, with the latter providing justification for the former. Here, we present an example of the knowledge network around the topic of herb-inspired drug discovery for AD.

Example. (Knowledge Network). As illustrated in Figure 10.3, the Hypothesis ":HAforAD," whose author is "Tong," states that the Herb ":Huperzine A" and the Disease ":Alzheimer's disease" seem to have the "treats" relationship. To prove the hypothesis, the Evidence ":Evidence," which is formed by SAs drawn from multiple data sources (each of which is represented as an RDF Graph), is proposed. The Hypothesis ":HAforAD" and the Evidence ":Evidence" are connected by

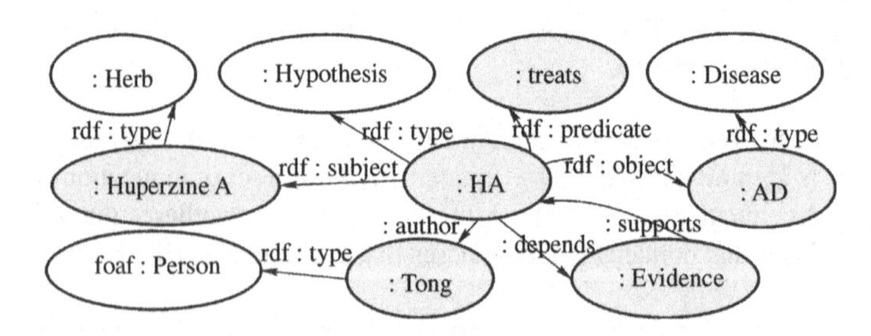

Figure 10.3 The evidentiary graph representing the hypothesis that "Huperzine A seems to treat Alzheimer's disease."

":supports" and "depends" relationships. The ":Evidence" itself may contain open hypotheses such as "Memory Loss is associated with Kidney Yang deficiency," which set goals for further investigation.

10.3.4 Semantic Schema

Agents need to know the global structure of the knowledge network and each other's content and interests in order to conduct effective communication. Here, we use the RDF Schema (RDFS) language [24] to define (1) the Global Hypothetical Graph (GHG) that specifies the structure and boundary of the knowledge network within a domain; (2) the Local Hypothetical Graph (LHG) that specifies an agent's hypothesis-answering capacity.

The RDF Schema defines two major graph structures [20,44]. First, given an ontology O that defines a set of classes C and the subsumption relationships (represented as the rdfs:subClassOf) between them, the inheritance Graph IG is a directed, acyclic graph (C, E) where C is the vertices set and E satisfies

$$(\forall c_1, c_2 \in C)[(c_1, \text{rdfs:subClassOf}, c_2) \Leftrightarrow c_2 \xrightarrow{p} c_1 \in E]$$

Second, the Property Graph PG represents the domain and range of properties. Given an ontology O that defines a set C of classes and a set P of properties with their domains and ranges, the property graph PG is a directed, labeled graph (C, E) where C is a set of vertices and E is such an edge set that

$$(\forall p \in P)[\{(p, \text{rdfs:domain}, d), (p, \text{rdfs:range}, r)\} \Leftrightarrow d \xrightarrow{p} r \in E]$$

We define the GHG as containing all the valid SA hypotheses specified by a Schema (IG, PG).

Definition 10.7. (GHG). The GHG specifies whether an SA hypothesis is valid in the following condition: GHG contains an hypothesis $?(s, p, o)$, iff class S and O exist, such that s and o are instances of S and O, respectively, and p has the domain D and range R that subsumes S and O, respectively.

The GHG defines a virtual information space which includes all that can be said about a domain, but contains no actual triples. For example, we can use the FOAF (Friends Of A Friend) ontology to define a global schema for the social network which contains the IG and PG as follows:

$$IG: \text{Person} \xrightarrow{\text{isa}} \text{Agent} \xrightarrow{\text{isa}} \text{Thing}$$
$$PG: \text{Person} \xrightarrow{\text{marriages}} \text{Person}$$

We can determine whether a hypothesis is legitimate for the social network. For example, if Mary and Jack are two persons, and Maggie is a dog, then "Mary marries Jack" is valid (even if Mary actually does not marry Jack) and the "Mary knows Maggie" is not. The actual social network can be conceptualized as a subset of the GHG that contains all legitimate hypotheses.

We move on to define the LHG as a subset of the GHG that specifies an agent's boundary of hypothesis-answering capacity.

Definition 10.8. (LHG). Given an ontology O that defines a set C of classes and a set P of properties, a Local Schema LS is a set of Atomic Views, each of which is represented by a hypothesis $?(a, p, b)$ such that a, b in C, p in P, and

$$a \xrightarrow{\text{isa}} d \xrightarrow{p} r \xleftarrow{\text{isa}} b.$$

An Atomic View $?(A, p, B)$ corresponds to a relation $R = \{(a, b)|a \text{ in } A, b \text{ in } B, (a, b) \text{ in } P\}$. The LHG contains all the valid hypotheses w.r.t. the Local Schema LS:

$$(a \xrightarrow{?p} b \in \text{LHG}) \Leftrightarrow (\forall V)[(V \in LS) \wedge (V \subseteq P) \wedge (a, b) \in V].$$

We let every agent maintain a Local Schema to express the capacity (or interests) of hypothesis answering. We assume that an agent can actually know less than that it claims in its local schema, and answer a subset of its LHG. Therefore, we define an Atomic View as a hypothesis.

For example, we can use a global schema (:Person, :knows, :Person) to define a global social network. An agent for Google, who might only maintain the social network of the employees of Google, can define the local schema as: (:Googler, :knows, :Googler), which is legitimate in that a Googler is a person who works for Google.

By using Semantic Schemas, we can declare an agent's capacity without revealing the physical structure of data resources maintained by the agent. This approach can facilitate query delegation or notification while preserving the access control of data.

10.3.5 Semantic Association Mining

We move on to discuss how a community of agents can find evidence for SA hypotheses from semantic graphs in a collaborative manner. We propose a recursive problem division strategy, in which one agent attempts to solve a published hypothesis with evidence that might depend on other hypotheses, which might already be proved or need to be further solved. In this strategy, an agent performs the following process recursively:

1. Select a new hypothesis to solve from a pool of published hypotheses;
2. Extract a pattern (as a subgraph) from its factual graph that can support the hypothesis;
3. Formalize the (possible) gap between the pattern (as what is proved) and the hypothesis (as what needs to be proved) as a set of hypothesis;
4. Publish the pattern as an evidence and possibly the newly derived hypotheses.

Through this strategy, a flow of knowledge around the original problem starts to spread through the network of agents. The agent can navigate in the knowledge flow to find a satisfying evidence for the hypothesis.

We first use an example in Figure 10.4 to illustrate this process. Suppose we are given a hypothesis: $H_1 = ?(1, p, 2)$, and we have three agents to collaborate in producing the evidence for the hypothesis. First, we assign the hypothesis to Agent A, who then discovers an evidence: $E_1 = \{(1, f, 3), ?(3, p, 5), (5, g, 2)\}$, which is partial in that it depends on a hypothesis $H_2 = ?(3, p, 5)$. Agent A publishes the evidence together with its dependency as: $\{(E_1, \text{depends}, H_1), (E_1, \text{supports}, H_2)\}$. Agent B provides an evidence $E_2 = \{(3, h, 4), (4, p, 5)\}$ for H_2, which also depends on $H_3 = ?(4, p, 5)$. Agent C provides a complete evidence $E_3 = \{(4, h, 6), (6, a, 5)\}$ for H_3. Through the collaboration between agents, we have a complete evidential graph for the original hypothesis H_1:

$$H_1 \xrightarrow{\text{dep}} E_1 \xrightarrow{\text{dep}} H_2 \xrightarrow{\text{dep}} E_2 \xrightarrow{\text{dep}} H_3 \xrightarrow{\text{dep}} E_3$$

Because the knowledge elements are distributed among three agents, an intelligent crawler is needed to recursively crawl evidence for hypotheses that it concerns, until complete evidence is found:

$$1 \xrightarrow{f} 3 \xrightarrow{h} 4 \xrightarrow{h} 6 \xrightarrow{a} 5 \xrightarrow{g} 2$$

In our approach, the SAM is carried out through the collaboration of different agents. Given a hypothesis $?(a, p, b)$, the generic strategy for an agent is:

1. Return SA (a, p, b) as direct primary evidence if available.
2. Propose a path P: $(r_0, p_1, r_1, p_2, r_2, \ldots, p_n, r_n)$, where $r_0 = a$, $r_n = b$.
3. For each subproblem $?(r_i, p_{i+1}, r_{i+1})$, replace it with an available SA(r_i, p_{i+1}, r_{i+1}) if possible.
4. Return P as new evidence.
5. Publish the newly generated hypotheses within P.
6. Go to Step 2 if more paths exist.

In this process, the basic task of probabilistic decision-making (PDM) for agents to perform is to substitute the SA Hypothesis $?(a, p, b)$ with a primary evidence (a, p, a_0) and a new hypothesis $?(a_0, p, b)$. Given a set of candidate primary evidences with a as the subject, the decision of choosing $?(a, p, a')$ over others is based on the expected complexity of solving $?(a', p, b)$.

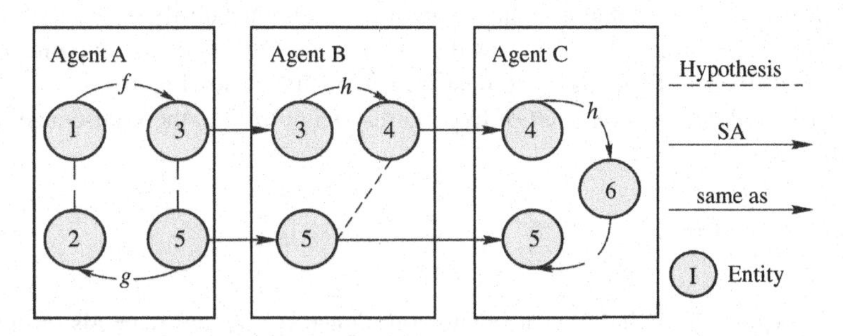

Figure 10.4 An example for Semantic Graph Routing.

10.3.6 Semantic Association Ranking

Because the number of relationships between entities is expected to outnumber entities themselves, SAM would more likely produce an overwhelming number of results for users than entity retrieval techniques, which brings more challenges to ranking schemes [45]. In our formulation, hypothesis ranking and evidence ranking are two distinct and interrelated problems:

1. The ranking of hypotheses based on their importance to decide which hypothesis to solve first.
2. The ranking of candidate evidence for one hypothesis based on the confidence of evidence.

We focus on collaborative ranking, in which ranking results shall emerge from the synergy of decentralized decision-making by a multitude of agents. The scheme shall fulfill the following goals: (1) the importance of hypotheses shall be estimated based on its potential utility and cost, so that agents could prioritize hypotheses to answer; (2) the topology of the evidentiary network should reflect the importance of hypotheses and evidence, and therefore graph-based ranking algorithms can be applied to select important hypotheses and evidence; (3) it shall be simple and scalable to a distributed environment. Accordingly, we propose a transitivity ranking mechanism based on the notions of confidence and utility. We suppose that each hypothesis can be assigned with an initial confidence and an initial utility upon generation, and then these weights can be updated according to the weights of their neighbors in the evidentiary network.

In our approach, hypotheses have two fundamental weights that guide agents to make decisions on their relative importance: (1) the confidence reflecting the expected cost of justification; (2) the utility reflecting the expected benefits of justification. First, each hypothesis is associated with a probabilistic truth value named Confidence C [0,1], so that for three particular forms of hypotheses, we have the following Confidence values:

1. $C = 0$ for absolutely impossible;
2. $C = 1$ for absolutely true;
3. $C = 0.5$ for absolutely unknown.

Second, we can also associate a hypothesis with a utility value U, which can be any positive value such that a higher value means a higher utility of the hypothesis. Therefore, we introduce $I = C \times U$ as the Importance Weight of the hypothesis.

The Confidence Transitivity is determined by the following formulae. For an evidence e that depends on a set of hypotheses $HG = \{h_i\}_{i=1}^{n}$, the confidence of e is determined by

$$C(e) = \prod_{i=1}^{n} C(h_i). \qquad (10.1)$$

For a hypothesis h that has an innate confidence $IC(h)$, and depends on a set of evidences $E = \{e_i\}_{i=1}^{n}$, the Confidence of h is determined by

$$C(h) = \max\left(IC(h), \{C(e_i)\}_{i=1}^{n}\right). \tag{10.2}$$

The Utility Transitivity is determined by the following formulae. For an evidence e that supports a set of hypotheses $HG = \{h_i\}_{i=1}^{n}$, the Utility of e is determined by

$$U(e) = \sum_{i=1}^{n} U(h_i). \tag{10.3}$$

For a hypothesis h that has an innate Utility $IU(h)$, and supports a set of evidences $E = \{e_i\}_{i=1}^{n}$, the Utility of h is determined by

$$U(h) = IU(h) + \sum_{i=1}^{n} U(e_i)/HN(e_i) \tag{10.4}$$

where $HN(e_i)$ is the number of hypotheses that support e_i.

10.3.7 Summary

In our mechanism, hypothesis generation and validation can give rise to a Hypothetical Graph, which is a set of entities connected by hypothetical relationships. Each edge corresponds to a hypothetical relationship and can be annotated by a series of evidences. A hypothetical network is incomplete, in that some edges are inadequately investigated or not annotated with sufficient evidences, and networks "grow" by making new hypotheses or providing evidences to prove or falsify the existing hypothesis. The current web, as the web of documents, allows people to share hypotheses and evidence in the form of hypertext or hypermedia. One scientist, if he/she has enough time, can search for and download a collection of documents around a topic and map out the evidentiary citations manually. However, manual processes are often inefficient and error-prone compared with machines in such kinds of mechanical tasks, and we will further evaluate the feasibility and usefulness of machines to understand and manipulate SAs on the Semantic Web of Data.

10.4 Evaluation

In this section, we establish a multiagent environment to simulate the situation of the Semantic Web and evaluate the feasibility and scalability of hypothesis-driven SAM through multiagent collaboration. The simulation has the following goals:

1. Investigate how our local evidence discovery strategy could affect global results.
2. Test the scalability of our collaboration strategy.
3. Investigate the characteristics of the resulting knowledge network.

10.4.1 Synthetic Graph Generation

We synthesize a set of graphs, each of which contains the following components:

1. The schema base (SB) contains a set of global schemas (each corresponds to a shared domain ontology that defines the relationships among classes and properties), each of which contains an Inheritance Graph and a Property Graph.
2. The resource list (RL) contains a multitude of resources, which might have types and properties for agents to calculate the similarity between resources.
3. The similarity matrix (SM) captures the pairwise similarity between resources.
4. The factual graph (FG) is the global graph of resources that contains all the SAs under investigation. We will tune the graph generation algorithm to investigate how the characteristics of the FG, e.g., size, homophily, community structure, and edge distribution, would affect the performance.
5. The Scalability Parameter K is the number of agents among which the graph is distributed. We use clustering methods to divide the FG into K interrelated communities, which are hosted by K interlinked agents, and analyze the performance with respect to K.

10.4.2 Engine Implementation

The Engine facilitates the collaboration through the registration of local schemas, the registration and notification of hypotheses and evidence, and the generation of complete proofs for hypotheses. Whereas multiple approaches exist for agents to communicate their capacity and knowledge elements, we choose to use a directory to enable the subscription and notification mechanism.

Directory: The engine contains a directory to maintain a global schema for a domain, and agents can register their local schema on the directory:

1. The global schema is maintained by the directory and is publicly accessible.
2. An agent A can register its local schema to the directory. The registration of an atomic view $V(u) = ?(a, p, b)$ created by A to the global schema is achieved by merging the hypothetical graph $?(a, p, b)$ to the Property Graph. Such a graph should be annotated with its provenance.
3. The resource list of an agent A can be crawled and indexed in the directory.
4. A Routing Table, the possibility of an agent A reaching resource o is measured by $P = \max_i \textit{Similarity}\ (a_i \in \mathrm{A}, o)$. In particular, if o in A, then $P = 1$. For a set of agents and a set of resources, the matrix $\mathrm{M}[n][m] = (P_i, j)$ is called the probabilistic routing table.

Notifier: The Notifier notifies a hypothesis $h = ?(s, p, o)$ to the agent A that is chosen from a set of candidates by the following criteria:

1. A has not been notified of h;
2. h belongs to A's Local Schema;
3. s belongs to A's Resource List;
4. A has the maximum possibility of reaching o (by consulting the Routing Table) among candidates that satisfy criteria 1, 2, and 3.

The candidates for global notification include all agents; and the candidates for local notification include neighbors of h's creator.

Proof Generator: As specified in Algorithm 10.1, the Proof Generator is an intelligent agent that works on the Evidentiary Graph in search of an optimized set of proofs for a hypothesis or hypothetical graph specified by users. It includes a Checker and a Crawler.

Algorithm 10.1 Proof Generation

```
Checker(e∈Evidence)
1:  for all h --dep--> e do
2:    setTrue (h.closed);
3:    for all (e' --dep--> h) ^ (isClosed(e') = True) do
4:      Checker(e');
5:    end for
6:  end for
isClosed(e∈Evidence)
1:  if(e.closed = False) ^ ( (∀e --dep--> h) h.closed = True)
    then
2:    setTrue(e.closed);
3:  end if
4:  return e.closed;
Crawler(h^c∈Closed Hypothesis)
1:  eg = Graph();
2:  for all e^c, h'^c: h^c --dep--> e^c --dep--> h'^c do
3:    eg.add (h^c --dep--> e^c --dep--> h'^c);
4:    eg.add(Crawler(h'^c));
5:  end for
6:  return eg;
```

The Checker, a procedure triggered by the generation of an evidence *e* with no depending hypothesis, recursively checks if the hypotheses and evidence that (indirectly) depend on *e* are closed. Here, an evidence is closed if (1) it does not depend on any hypothesis or (2) it depends only on hypotheses that are closed; a hypothesis is closed if it is supported by closed evidence.

The Crawler, a procedure triggered by the client posing a hypothesis *h*, navigates on the global Evidentiary Graph to generate a subgraph that proves *h*.

10.4.3 Miner Implementation

The knowledge space contains a set of agents (called Miner) that perform evidence discovery, hypothesis generation, and their publication.

Local Knowledge Base: A Miner maintains a local knowledge base that contains the following components: (1) a shared domain ontology with local schemas; (2) a resource list; (3) semantic graphs for expressing facts; and (4) published hypotheses and evidence.

Router: Evidence Discovery is implemented as Semantic Graph Routing (see Algorithm 10.2). Within each iteration of the recursive routing procedure (shown as follows):

$$s \overset{t}{\dashrightarrow} c \overset{p}{\longrightarrow} j \overset{?h'}{\longrightarrow} o, \qquad s \overset{?h}{\longrightarrow} o$$

Algorithm 10.2 Semantic Graph Routing

```
Router(s, c, o∈Resource, t∈Graph)
1:  if c --?p--> o∈HG then
2:     return new Evidence (s --t--> c --?p--> o);
3:  end if
4:  if c --p--> o∈FG then
5:     t.add (c --p--> o);
6:     return new Evidence (s --t--> o);
7:  end if
8:  Resource j = Next(c, o, t);
9:  if (j = null) then
10:     new Hypothesis (c --?p--> o∈ HG);
11:     return new Evidence (s --t--> c --?p--> o);
12:  else
13:     t.add (c --p--> o);
14:     return Router(s, j, o, t);
15:  end if
Next(c, o∈Resource, t∈Graph)
1:  for all {j_i}_{i=1}^n (c→j_i∈FG) ^ (j_i∉t) do
2:     j: I(j, o) = max_{i=1}^n I(j_i, o);
3:  end for
4:  if (j≠null) ^ (I(j,o) ≥ I(c,0)) then
5:     return j;
6:  else
7:     return null;
8:  end if
```

The Router is given a hypothesis $?(s, h, o)$ as the original problem, and a trace (which is a path within FG) $t = (s, \ldots, c)$ as the current solution. It first chooses j (as the next "Jumping Point") from neighbors of c and updates the trace t as $(s, \ldots, j)$, and then invokes the next iteration with h and t.

The choice of j from neighbors of c is performed by the Next function, which optimizes the possibility of j associated with o with the following procedure:

Step 1: Reason if the FG entails (j, p, o).

Step 2: Calculate the estimated value $I \times S$, in which I is j's importance as measured by its degree, and S is the similarity between j and o.

The routing procedure terminates within an agent's local boundary under the following conditions:

Step 1: The remaining problem ?(*c*, *h*, *o*) is an identified hypothesis;
Step 2: $j = o$;
Step 3: *j* is null (*c* is the locally optimized point for *h*).

In addition, the routing procedure can be relayed to the agent's neighbors through local notification. If agent A fails to find such a neighboring agent to relay *h*, then it will invoke the global notification mechanism for *h*.

10.4.4 Collaborative Discovery Process

We first set up a simulation environment, and generate a set of hypotheses *H*, and then we run the simulation process for time *T*, and inspect the result.

- *Environment setup*: We generate a Synthetic Graph SG, an engine, and *K* agents, and assign agents with data. We let the engine and agents start listening.
- *Hypothesis generation*: The Original Hypothesis Generator Agent *hg* creates a set of seeding valid hypotheses. The engine then performs the global notification mechanism for the seeding hypotheses to initiate the evidence discovery process.
- *Evidence discovery*: The engine and agents perform evidence discovery through routing and notification until all seeding hypotheses are solved or the time *T* runs out.

We first present an example (Figure 10.5) to explain the process. The Factual Graph FG contains 21 vertices (of the same class), 72 edges (of the same property),

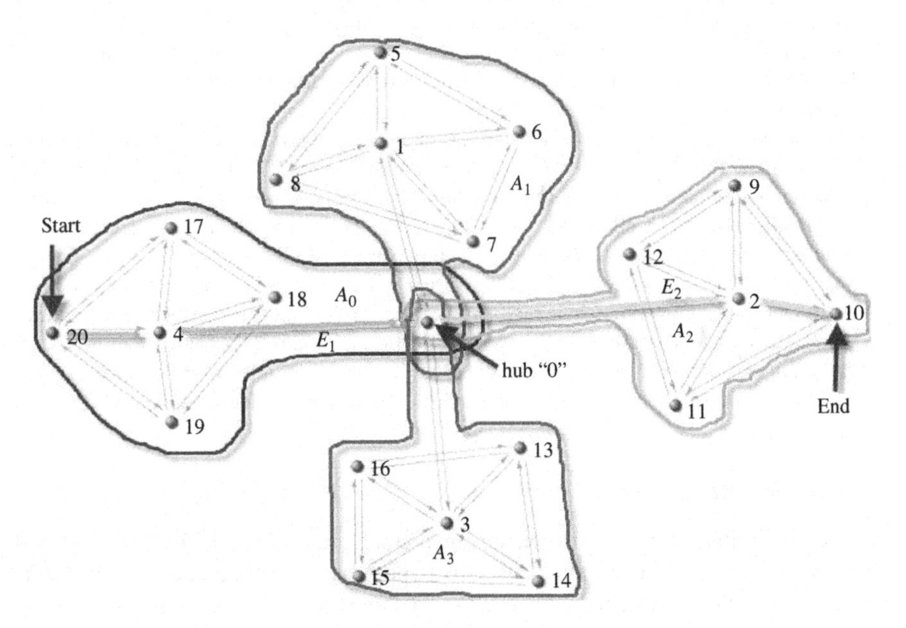

Figure 10.5 A setting where four agents hold four communities connected by the node 0.

and 4 obvious communities with the overlapping node as "0." The similarity between node i and node j is defined as $1/(D(i, j) + 1)$, where $D(i, j)$ is the distance between i and j (with $D(i, j) = 0$ for $i = j$). We assign each community to one agent and start the process P with a seeding hypothesis $h_1 = ?(20, p, 10)$:

1. The engine notifies Agent A_0 of h_1 because only Agent A_0 contains h_1's subject "20."
2. Agent A_0 routes through the path (20, p, 4, p, 0) to optimize the similarity between the current node and the target "10." Agent A_0 publishes the evidence $e_1 = (20, p, 4, p, 0, p, 10)$ which depends on $h_2 = ?(0, p, 10)$.
3. Agent A_2 is notified of h_2 because the possibility P of A_2 reaching "10" ($P = 1$ since A_2 contains) is the highest among the neighbors of A_0.
4. Agent A_2 routes through the path (0, p, 2, p, 10) based on the same strategy as A_0 and publishes the path as an evidence e_2 for h_2.
5. The publication of e_2 triggers the proof checker, which sets h_2, e_1, h_1 as closed.
6. Now that h_1 is closed, the discovery process is terminated.
7. Upon the client's request, the crawler can generate an evidentiary graph:

$$\{(h_1, \text{dep}, e_1), (e_1, \text{dep}, h_2), (h_2, \text{dep}, e_2)\}$$

which entails the evidence (20, p, 4, p, 0, 2, 10).

We compare P with an alternative process A, in which the same graph is only assigned to one agent, who finds the same final evidence by using the same strategy as Agent A_0, with the following measurements:

- *Coverage ratio c*: The coverage ratio c, as the ratio of traversed statements to all statements in FG, is a major factor for measuring agents' efficiency. Process P has $c = 4/72$ that is equal to the c of A, which means that P is efficient.
- *Average evidence size* $\overline{|e|}$: Suppose a process generates $\{e_i\}_{i=1}^n$, $\overline{|e|} = \frac{1}{n}\sum_{i=1}^n |e_i|$ is a critical factor for measuring agent's responsiveness. For P, the average evidence size is $(|e_1| + |e_2|)/2 = 2$; For A, the average evidence size is 4: through multiagent collaboration, the responsiveness of each agent increases.
- *Participation ratio w*: The ratio of working miners (who participate in the collaboration) to all miners. For P, the collaboration graph shows that $w = 2/4 = 0.5$; for A, $w = 1$.
- *Total computational time t*: For the engine E and miners $\{A_i\}_{i=1}^n$, the t is the sum of computational time of all agents: $t = T(E) + \sum_{i=1}^n T(A_i)$. The change of factor t over a different number of agents reflects the scalability of the mechanism.
- *Success ratio s and serendipitous discovery ratio d*: Suppose a process that is given a set S of seeding hypotheses, actually solves the set R of hypotheses. The success ratio $|S \cap R|/|S|$ and the serendipitous discovery ratio $|R - S|/|S|$ are two factors that determine the quality of mining. For P, $s = 1/1$ and $d = 1/1$; for A, $s = 1/1$ and $d = 0$, meaning that the multiagent collaboration has the advantage of discovering some extra knowledge.

10.4.5 Result Analysis

As shown in Figure 10.6, we generate a factual graph with 100 nodes and 861 edges and perform each iteration with 10 hypotheses. We examine the scalability of our approach by watching the change in the above factors as the number of agents K increases from 1, 2, 4, 6, 8, to 10. For each K, we perform the process for

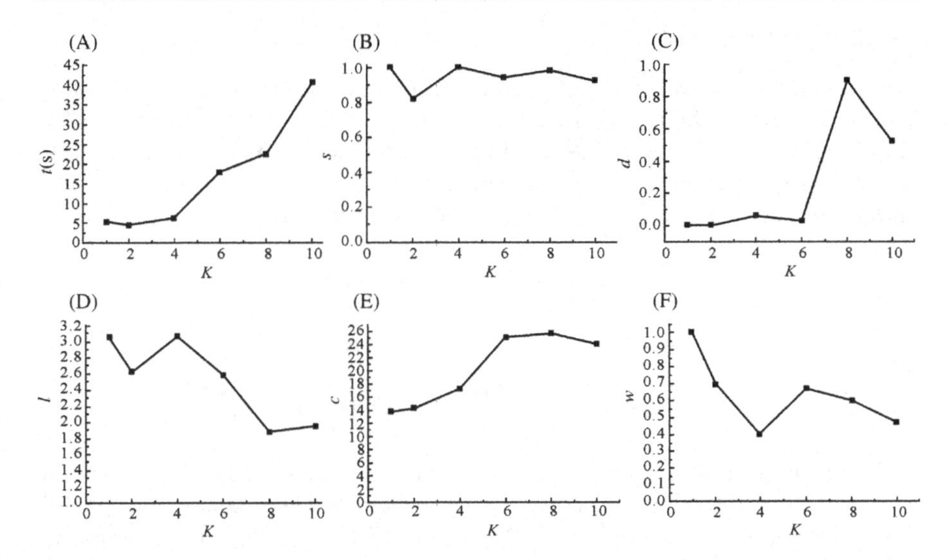

Figure 10.6 The results of a simulation process testing the scalability of the proposed mechanism.

5 iterations. Overall, we proposed 300 hypotheses and finally solved 283 of them with an overall success ratio of 94.3%, which shows that our approach is very effective. The 5.7% unsolved hypotheses indicate the fact that semantic similarity does not necessarily mean connectivity in a graph. Figure 10.6A shows that as the number of agents grows, the total computational time t increases linearly. Figure 10.6B shows that the success ratio s remains high as the number of agents grows. Figure 10.6C shows that the serendipitous discovery ratio d (as the ratio of closed derivative hypotheses to seeding hypotheses) increases as the number of agents grows. Figure 10.6D shows that as the number of agents grows, the average evidence size decreases, and hypotheses are more frequently solved by the combination of evidence. Figure 10.6E shows that as the number of agents grows, the coverage ratio c increases. Figure 10.6F shows that as the number of agents grows, the participation ratio w decreases. These results show that: (1) the proposed algorithm can scale up to a large number of agents while sustaining the quality and performance of discovery; (2) as the number of agents increases, serendipitous discovery can also increase.

10.5 Use Cases

In this section, we evaluate the effectiveness of our framework in integrative studies such as formulae system interpretation [10] and herb-inspired drug discovery [25]. We will first present an overview of the major use case—the herb network—and carry out the evaluation through a set of specific use cases, each of which includes the problem description, the main process, and example presentation and interpretation.

10.5.1 The HerbNet

The pharmaceutics domain mainly studies the complex interactions between drugs and components of the human body. In a mathematical language, the knowledge in this domain can be abstracted as a bipartite graph between drugs and targets [33]. For example, the knowledge that "Huperzine A is an inhibitor of the enzyme cholinesterase (a drug target for AD) [27,28]" can be abstracted as:

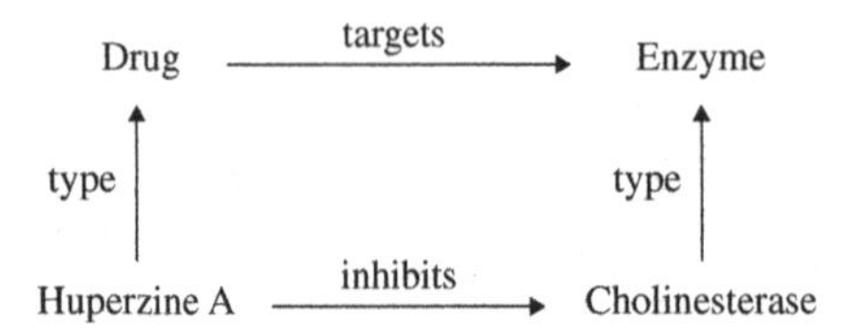

TCM physicians primarily prescribe a concoction of multiple herbs and establish a system of Herbal Formulae as significant patterns of herb composition [5,10]. For example, the Four-Gentlemen Decoction (FGD) consists of the following four herbs: (1) *Ginseng* [Gins]; (2) *Atractylodis* [Atr]; (3) *Sclerotium* [Scler]; (4) *Glycyrrhiza uralensis* [Glyc]. The formulae system gives rise to a bipartite graph between the formulae and herbs:

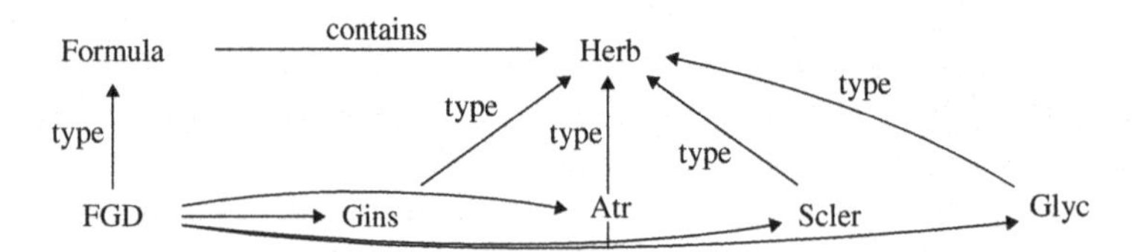

In addition, a formula can also be interpreted as a collaboration of its member herbs, which leads to an herb-collaboration graph:

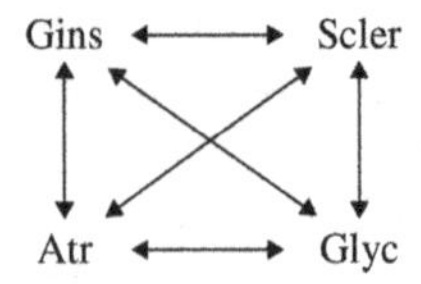

Herb-inspired drug discovery attempts to interpret the phenomena of formula concoction in terms of the underlying drug–drug interactions and/or drug–target interactions, in order to discover bioactive compounds from herbs or examine the efficacy and toxicity of formulae [25]. We aim to evaluate our system's capability to support the following requirements in herb-inspired drug discovery:

1. *Cross-disciplinary collaboration*: A seamlessly integrated approach to using the knowledge in one domain, to support the decision-making in another domain.
2. *Actionable knowledge*: The agents should be able to "understand" the problem-solving context of the user and provide valuable knowledge to support decision-making of the user.

3. *Serendipitous reuse*: The knowledge that is originally discovered or explicitly represented for the purpose of solving one problem is reused, after its publication, to solve another problem.
4. *Serendipitous discovery*: Discover valuable knowledge that is unexpected by the knowledge analyst at the beginning of the knowledge discovery process.
5. *Knowledge reliability*: The discovered knowledge should be faithfully represented (a formal representation can facilitate machine processing) and supported by reliable evidence.

10.5.2 Formula System Interpretation

Frequent patterns are itemsets, subsequences, or substructures that appear in a dataset with a frequency no less than a user-specified threshold [46]. For a knowledge base containing a set of graphs, the problem of Frequent Semantic Subgraph Discovery (FSSD) [37] is defined as follows:

Definition 10.9. (FSSD). Given a set S of graphs $\{G_i\}_{i=1}^n$, and the global graph $GG = \cup_{i=1}^n G_i$ as the merger of the graphs in S. A subgraph p of GG is a Frequent Semantic SubGraph (FSSG) of S, or simply frequent iff p occurs in the graphs of S no less than $\theta|S|$ times, where θ is a user-specified minimum support threshold, and $|S|$ is the number of graphs in S. Let $S_p = \{g \mid p \subseteq g,\ g \in S\}$ be all the supergraphs of p within S.

In our FSSD process, we first construct a set of named graphs using user-specified semantic queries, and then feed the resulting graph set to an existing frequent pattern mining operator which treats each statement as an item [47]. We put emphasis on mechanical interpretation of FP (Frequent Pattern) based on domain logics, which is an important yet unsolved issue. The problem of generating semantic annotations for FPs with context analysis is first introduced in Ref. [42]. In our approach, a discovered pattern is essentially evidence, which can be associated with other evidences via the evidentiary graph. Therefore, discovered patterns can be annotated with neighboring domain knowledge based on SAs and visualized as a rich graph to facilitate human interpretation.

Suppose a pharmaceutical analyst retrieves a set of clinical cases related to "Kidney Yang Deficiency (KYD)" [13] (an instance of TCM Syndrome) to discover interesting patterns related to drug usage, efficacy, and safety. The process is as follows:

1. Represent the content of clinical cases as a set of semantic graphs.
2. Merge these graphs with domain knowledge by adding annotations to every graph.
3. Run an FPD algorithm against the graph set, which results in a set of discovered patterns.
4. Add semantic annotations to every pattern based on SAM.
5. Visualize patterns for user interpretation.

An FSSD is conducted against the integrated dataset generated from a clinical warehouse, a Structured Literature repository, and a Chinese Medical Formula database, which results in a set of patterns. For example, Figure 10.7 showing a pattern including four herbs and two drug efficacies is interpreted by the fact that the formula FGD composed of these herbs has these two drug efficacies.

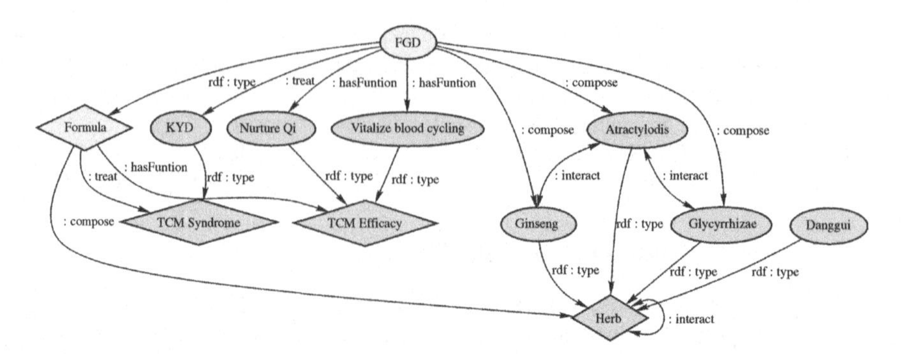

Figure 10.7 Discovered patterns can be annotated with domain knowledge based on SAs of concepts and visualized as a semantically enriched graph to facilitate human interpretation.

10.5.3 Herb–Drug Interaction Network Analysis

Network-based clustering is defined as an unsupervised partition of network nodes into a collection of communities, based on interaction density, or other related indicators such as node similarity (through sharing of properties) and relational strength [31].

Interactions between herbs and drugs are critical drug safety issues related to TCM [48]. By aggregating herb–drug interactions into a knowledge network, we can further analyze node centrality and community structure within the network [37].

Suppose we are given a knowledge base *KB*, a target class *D*, and a problem-solving context *PC* expressed in a *SG*. We first find all SAs between each pair of individuals of *D* from *KB* that are related to *PC*, and then group individuals of class *D* into communities, such that there is a higher density of interactions within communities than between them. The process is as follows:

1. Find all interesting SAs between individuals of *D* from *KB* (by generating and executing a set of SPARQL queries) related to *PC* and merge the resulting SAs into a semantic graph *SG*. Aggregate statements in *SG* w.r.t. *KB*, which transforms *SG* into a concise, weighted graph.
2. Add all components (a graph component refers to a connected subgraph) of *SG* into the set *CS*.
3. Recursively pick a component *c* in *CS* to test if *c* satisfies the criteria for a community.
4. Move *c* from *CS* to the result set *RS* if *c* is a community, and split *c* into smaller components otherwise. The splitting of *c* takes three major substeps: (1) remove centrality resources (whose closeness centrality ratio exceeds a user-defined value) from *c*; (2) recursively remove the statement with max betweenness centrality ratio until *c* ceases to be connected; (3) add centrality resources into every component of *c*. After *c* is divided into a set of components, add these components into *CS* and remove *c* from *CS*.
5. Generate semantic annotations for discovered communities with context analysis. For every community *c* in *RS*, ask *PC* to find all interesting SAs between individuals of *c* from *KB* (by generating and executing a set of SPARQL queries), and to merge the resulting SAs into *c*.

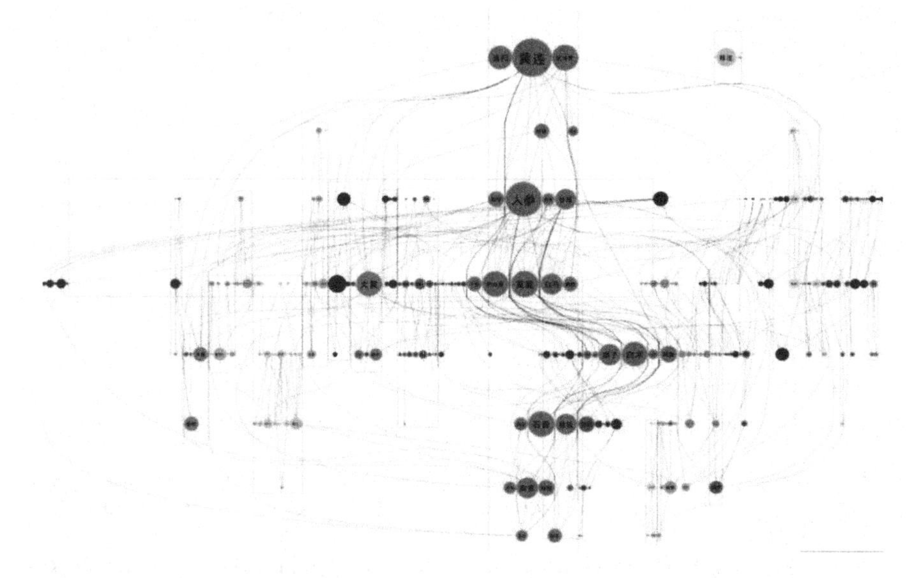

Figure 10.8 Global network of herb–drug interactions, with drugs represented by nodes with size/font proportional to degrees, interactions represented by edges, and drug communities represented by distinct colors.

In the resulting network (Figure 10.8), most nodes (99.3%) participate in the largest connected components, and there is also a big drug community (pink colored) that consists of many of the biggest hubs in the network, including the four herbs in FGD, and reveals that drug hubs tend to cluster together in the TCM domain. The clustering results are evaluated by domain experts as correctly reflecting the TCM clinical practice.

10.6 Conclusions

In this chapter, we proposed a hypothesis-driven framework for SA discovery in a distributed environment. The framework utilizes shared domain ontologies to facilitate a multitude of agents to communicate hypotheses and evidence about SAs. This decentralized collaboration can give rise to a knowledge network that synthesizes complementary knowledge elements discovered from distributed data. The simulation shows that the framework can scale up to a large number of agents and ensure the quality of discovered SAs. Through this mechanism, we utilize the TCM ontology to discover and integrate relationships and interactions centered on herbal medicine from distributed data sources, resulting in a knowledge network that supports such knowledge discovery tasks as frequent pattern discovery from herbal formulae and herb–drug interaction network analysis. The case study shows that our framework and the Semantic Web technology are promising in collaborative D^3M across disciplinary boundaries.

References

[1] Han J, Kamber M. Data mining: concepts and techniques. San Francisco, CA: Morgan Kaufmann Publishers; 2006.

[2] Cao L, Zhang C, Liu J. Ontology-based integration of business intelligence. Int J Web Intell Agent Syst 2006;4(4):1–14.

[3] Cao L, et al. Intelligence metasynthesis in building business intelligence systems, WImBI 2006, LNAI 4845; 2007. p. 454–70.

[4] Cao L, Zhang C. Domain-driven, actionable knowledge discovery. IEEE Intell Syst 2007;22(4):78–88.

[5] Patwardhan B, Warude D, Pushpangadan P, et al. Ayurveda and traditional Chinese medicine: a comparative overview. Evid Based Complement Alternat Med 2005;2:465–73.

[6] Rees L, Weil A. Integrated medicine: imbues orthodox medicine with the values of complementary medicine. BMJ 2001;322(7279):119–20.

[7] Wisneski L, Anderson L. The scientific basis of integrative medicine. Boca Raton, FL: CRC Press; 2004.

[8] Mao Y, et al. Dynamic sub-ontology evolution for traditional Chinese medicine web ontology. J Biomed Inf 2008;41(5):790–805.

[9] Chen H, et al. From legacy relational databases to the Semantic Web: an in-use application for traditional Chinese medicine. Proceedings of the 5th international Semantic Web conference; 2006. p. 109–20.

[10] Yi Y, Chang I. An overview of traditional Chinese herbal formulae and a proposal of a new code system for expressing the formula titles. Evid Based Complement Alternat Med 2004;1:125–32.

[11] Yang AW, Allan G, Li CG. Effective application of knowledge management in evidence-based Chinese medicine: a case study. Evid Based Complement Alternat Med 2009; 6(3): (September)393–8.

[12] Feng Y, et al. Knowledge discovery in traditional Chinese medicine: state of the art and perspectives. Artif Intell Med 2006;38(3):219–36.

[13] Zhou X, Liu B, Wu Z. Integrative mining of traditional Chinese medicine literature and MEDLINE for functional gene networks. Artif Intell Med 2007;41:87–104.

[14] Lee F, Herman I, Hongsermeier T, et al. The Semantic Web in action. Sci Am 2007;297:90–7.

[15] Ruttenberg A, et al. Advancing translational research with the Semantic Web. BMC Bioinformatics 2007;8(Suppl. 3):S2.

[16] Rubin D, Shah N, Noy N. Biomedical ontologies: a functional perspective. Brief Bioinform 2008;9:75–90.

[17] Ciccarese P, et al. The SWAN biomedical discourse ontology. J Biomed Inform 2008;41(5):739–51.

[18] Clark T, Kinoshita J. Alzforum and SWAN: the present and future of scientific web communities. Brief Bioinform 2007;8(3):163–71.

[19] Gao Y, et al. SWAN: a distributed knowledge infrastructure for Alzheimer disease research. J Web Sem 2006;4(3):222–8.
[20] Mukherjea S, Bamba B, Kankar P. Information retrieval and knowledge discovery utilizing a biomedical patent Semantic Web. IEEE Trans Knowl Data Eng 2007;19 (2):1099–110.
[21] Mukherjea S. Information retrieval and knowledge discovery utilizing a biomedical Semantic Web. Brief Bioinform 2005;6:252–62.
[22] Anyanwu K, Sheth A. P-queries: enabling querying for semantic associations on the Semantic Web. Proceedings of the 12th international conference on World-Wide Web; 2003. p. 20–4.
[23] Anyanwu K. Supporting link analysis using advanced querying methods on Semantic Web datasets, Ph.D. Dissertation; 2007.
[24] Beckett D. 2007. RDF/XML syntax specification, w3c rec, <http://www.w3.org/TR/rdf-syntax-grammar/>; 2009.
[25] Fabricant D, Farnsworth N. The value of plants used in traditional medicine for drug discovery. Environ Health Perspect 2001;109(1):69–75.
[26] Xiao X, Wang R, Tang X. Huperzine A and tacrine attenuate beta amyloid peptide-induced oxidative injury. J Neurosci Res 2000;61(5):564–9.
[27] Perry E, et al. Medicinal plants and Alzheimer's disease: integrating ethnobotanical and contemporary scientific evidence. J Altern Complement Med 1998;4(4):419–28.
[28] Gaeddert A. The herbal approaches to treating memory loss, dementia, and Alzheimer's disease. Acupunct Today 2003;4(4).
[29] Selkoe DJ. Alzheimer's disease: genes, proteins, and therapy. Physiol Rev 2001;81(2):41–66.
[30] Berners-Lee T, et al. Tabulator redux: browsing and writing linked data. Proceedings of WWW workshops: Linked Data on the Web (LDOW); 2008.
[31] Chakrabarti D, Faloutsos C. Graph mining: laws, generators, and algorithms. ACM Comput Surv 2006;38(1):9–32.
[32] Barabasi A, Oltvai Z. Network biology: understanding the cells functional organization. Nat Rev Genet 2004;5:101–13.
[33] Barabasi A. Network medicine from obesity to the diseasome. N Engl J Med 2007;357:404–7.
[34] Fang YC, Huang HC, Chen HH, et al. TCMGeneDIT: a database for associated traditional Chinese medicine, gene and disease information using text mining. BMC Complement Altern Med 2008;8(1):58.
[35] Cohen A, Hersh W. A survey of current work in biomedical text mining. Brief Bioinform 2005;6(1):57–71.
[36] Jiang T, Tan A, Wang K. Mining generalized associations of semantic relations from textual web content. IEEE Trans Knowl Data Eng 2007;19(2):164–79.
[37] Chen H, et al. Semantic Web for integrated network analysis in biomedicine. Brief Bioinform 2009;10(2):177–92.
[38] Bizer C, Cyganiak R, Heath T. How to publish linked data on the web, <http://sites.wiwiss.fu-berlin.de/suhl/bizer/pub/LinkedDataTutorial/>.
[39] Berners-Lee T, Hall W, Hendler J, et al. A framework for web science. Found Trends Web Sci 2006;1(1):1–130.
[40] Gutierrez C, Hurtado C, Mendelzon A. Foundations of Semantic Web databases. Proceedings of the 23th PODS; 2004. p. 95–106.
[41] Carroll J, et al. Named graphs, provenance and trust. Proceedings of the 14th international conference on World-Wide Web; 2005. p. 613–22.

[42] Mei Q, et al. Generating semantic annotations for frequent patterns with context analysis. Proceedings of the 12th ACM SIGKDD; 2006. p. 337–46.
[43] Aleman-Meza B, et al. Semantic analytics on social networks: experiences in addressing the problem of conflict of interest detection. Proceedings of the 15th international conference on World-Wide Web; 2006. p. 407–16.
[44] Theoharis Y, Tzitzikas Y, Kotzinos D, et al. On graph features of Semantic Web schemas. IEEE Trans Knowl Data Eng 2008;20(5):692–702.
[45] Aleman-Meza B. Ranking complex relationships on the Semantic Web. IEEE Int Comput 2005;9(3):37–44.
[46] Han J, Cheng H, Xin D, et al. Frequent pattern mining: current status and future directions. Data Min Knowl Discov 2007;15(1):55–86.
[47] Wu Z, et al. Semantic Web development for traditional Chinese medicine. Proceedings of AAAI-08/IAAI-08; 2008. p. 1757–62.
[48] Fugh-Berman A. Herb–drug interactions. Lancet 2000;355:134–8.

11 Semantic-Based Database Integration for Traditional Chinese Medicine[1]

11.1 Introduction

Up to now, many killer applications reported by the Semantic Web community often focus on processing the unstructured document data, using semantic annotation or various learning, mining, and natural language processing techniques [1]. However, data in big organizations are normally stored in relational databases or other appropriately formatted documents. Over-emphasizing those applications, which handle unstructured documents, may obscure the community from the fact that the essence of the Semantic Web comes from its similarity to a huge distributed database. To back up this idea, consider the following statements made by Tim Berners-Lee in 2005 about his vision of the future Semantic Web:

> The Semantic Web is not about the meaning of documents. It's not about marking up existing HTML documents to let a computer understand what they say. It's not about the artificial intelligence areas of machine learning or natural language understanding... It is about the data which currently is in relational databases, XML documents, spreadsheets, and proprietary format data files, and all of which would be useful to have access to as one huge database

From this point of view, one of the ways to realize the vision of the Semantic Web is (1) to interconnect distributed located legacy databases using richer semantics; (2) to provide ontology-based query, search, and navigation as one huge distributed database; and (3) to add additional deductive capabilities on the top to increase the usability and reusability of data.

We have developed and deployed such a kind of Semantic Web application for the China Academy of Traditional Chinese Medicine (CATCM). It semantically interconnects over 70 legacy traditional Chinese medicine (TCM) databases by a formal TCM ontology with over 70 classes and 800 properties. As illustrated in Figure 11.1, the basic idea is to construct a separate semantic layer to fill up the

[1] Chen H, Wang Y, Wang H, Mao Y, Tang J, Zhou C, Yin A, Wu Z. Towards a semantic web of relational databases: a practical semantic toolkit and an in-use case from traditional Chinese medicine, ISWC 2006;4273:750–63. With kind permission from Springer Science + Business Media.

Modern Computational Approaches To Traditional Chinese Medicine. DOI: http://dx.doi.org/10.1016/B978-0-12-398510-1.00011-X

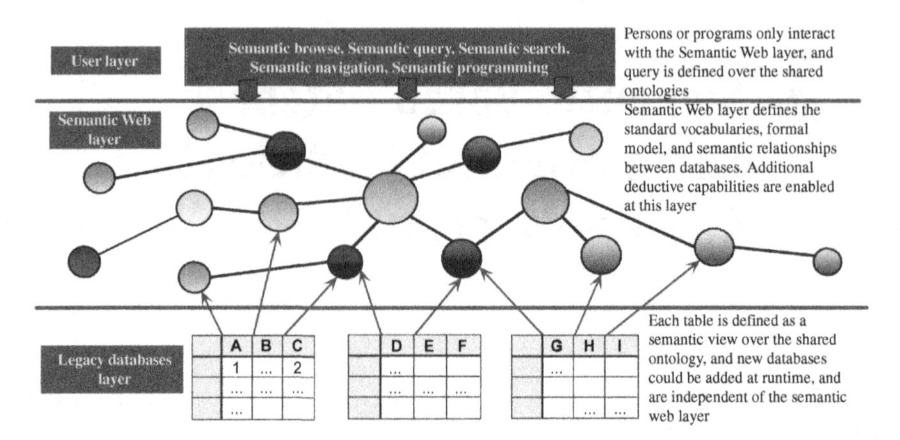

Figure 11.1 Toward a Semantic Web of relational databases.

gaps among legacy databases with heterogeneous structures, which might be semantically interconnected. Users and machines only need to interact with the semantic layer, and the semantic interconnections allow them to start in one database and then move around an extendable set of databases. The semantic layer also enables the system to answer semantic queries across several databases such as "What diseases does this drug treat?" or "What kind of drugs can treat this disease?"; not like the keyword-based searching mechanism provided by conventional search engines.

This application is built upon a framework called DartGrid. A set of well-implemented semantic tools of DartGrid are introduced in this chapter. In particular, a visualized mapping tool is developed to facilitate domain experts to define semantic mapping from different relational schemas to the shared ontology; the query interface is dynamically generated based on class definitions and could construct a SPARQL query which will be rewritten into a set of SQL queries; an intuitive search interface is provided to enables users to select the most appropriate and accurate results possible by a concepts ranking mechanism.

The chapter is organized as follows. Section 11.2 talks about the system architecture and technical features. Section 11.3 elaborates on the implementation of the semantic mediator and the visualized semantic mapping tool. Section 11.4 introduces the TCM semantic portals which provide semantic query and search services. Section 11.5 reports the user evaluation and lessons learned from this developing life cycle. Section 11.6 mentions some related work. Section 11.7 gives a summary and our future directions.

Please also note that due to the special character of TCM research, in which the terminologies and definitions are not always interpretable, some figures in this chapter contain Chinese search results and web interface. We have annotated all the necessary parts of the figures in English, and we would expect it would be sufficient to understand the functionalities of this application.

11.2 System Architecture and Technical Features

11.2.1 System Architecture

The system applies service-oriented architecture (SOA) and several web service technologies. As Figure 11.2 depicted, four key components are implemented as web services in the core.

1. *Ontology service* is used to expose the shared ontologies that are defined using web ontology languages (OWL). Typically, the ontology is specified by a domain expert who also is in charge of the publishing, revision, and extension of the ontology.
2. *Semantic registration service* maintains the semantic mapping information. Typically, database providers define the mappings from relational schema to domain ontology and submit the registration entry to this service.
3. *Semantic query service* is used to process SPARQL semantic queries. First, it gets mapping information from the semantic registration service. Afterward, it translates the semantic queries into a set of SQL queries and dispatches them into specific databases. Finally, the results of SQL queries will be merged and transformed back to semantically enriched format.
4. *Search service* supports full-text search in all databases. The search results will be statistically calculated to yield a concepts ranking, which help the user to get more appropriate and accurate results.

11.2.2 Technical Features

The following are five features that distinguish this application from other similar semantic data integration tools, which will be introduced in detail in Section 11.6.

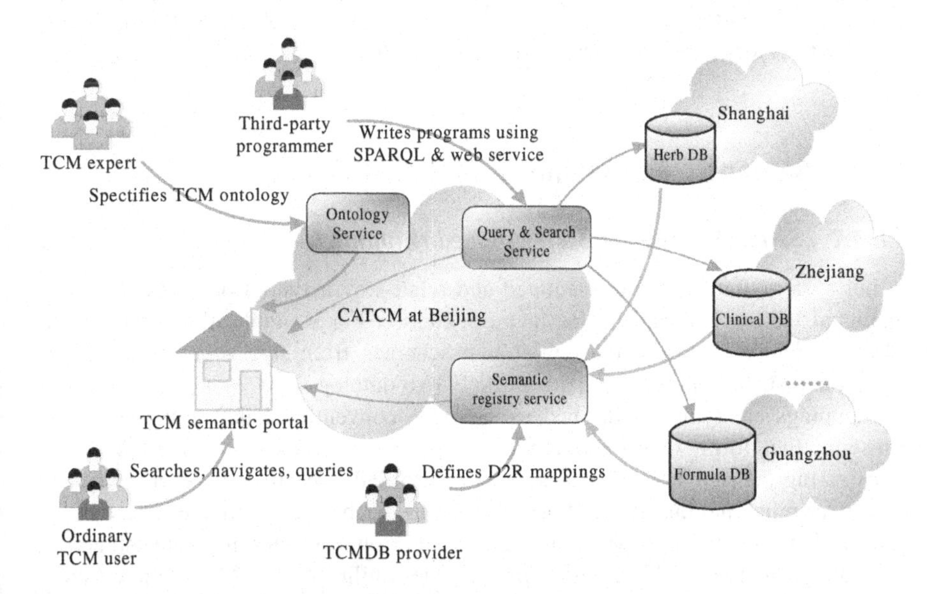

Figure 11.2 SOA-based system architecture and usage scenario.

- *View-based semantic mapping.* In our system, an ontology acts as the semantic mediator for heterogeneous databases. Relational database schemas are mapped into corresponding classes or properties and related by a semantic relationship defined in this ontology. To be specific, the mappings are defined as semantic views, i.e., each relational table is defined as a view over this shared ontology. According to conventional data integration literature, a view-based approach has a well-understood foundation and has been proved to be flexible for heterogeneous data integration.
- *View-based query rewriting with additional inference capabilities.* A view-based query rewriting algorithm is implemented to rewrite the SPARQL queries into a set of SQL queries. This algorithm extends earlier relational and XML techniques for rewriting queries using views, with consideration of the features of OWL. Otherwise, this algorithm is enriched also by additional inference capabilities on predicates such as subClassOf and subPropertyOf.
- *Visualized semantic mapping tool.* Defining mappings is a labor-intensive and error-prone task. In our system, a new database could be added to the system at runtime by using a visualized mapping tool. It provides many ease-of-use functionalities such as drag-and-drop mapping, mapping visualization, data source annotation, and so on, to facilitate the user in this task.
- *Dynamic query interface.* A form-based query interface is offered to construct semantic queries over shared ontologies. It is automatically generated at runtime according to property definitions of classes and will finally generate a SPARQL query.
- *Intuitive search interface.* This Google-like search interface accepts one or more keywords and makes a complete full-text search in all databases. Users could semantically navigate in the search results and move around an extendable set of databases based on the semantic relationships defined in the semantic layer. Meanwhile, the search system could generate a suggested list of concepts which are ranked based on their relevance to the keywords. Thereafter, users could explore the semantic query interface of those concepts, and specify a semantic query in them to get more accurate and appropriate information. We call it an intuitive search because it could generate a list of concept suggestions to help the user improve the search results.

11.3 Semantic Mediation

11.3.1 Semantic View and View-Based Mapping

In our system, databases are mediated and related by a shared ontology, and each relational table is mapped in one or more classes. For example, the mapping scenario in Figure 11.3 illustrates relational schemas from two sources (W3C and ZJU) and a shared ontology (a part of the FOAF ontology).

Mappings are typically defined as views in conventional data integration systems in the form of global-as-view (GAV) and local-as-view (LAV) [2]. Considering the Semantic Web case, GAV is to define each class or property as a view over relational tables, and LAV is to define each relational table as a view (or query) of the shared ontology. The experiences from conventional data integration systems tell us that LAV provides greater extensibility than GAV: the addition of new sources is less likely to require a change to the mediated schema [2]. In our

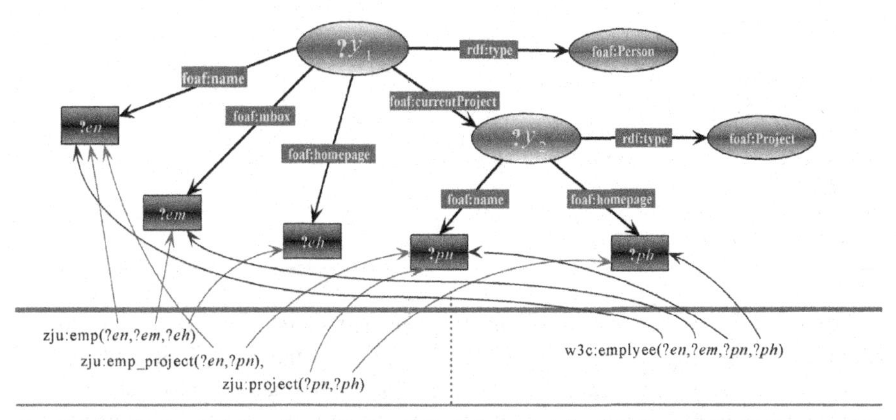

View-1:zju:emp(?*en*,?*em*,?*eh*):-($?y_1$,rdf:type,foaf:Person),($?y_1$,foaf:name,?*en*),($?y_1$,foaf:mbox,?*em*),($?y_1$,foaf:homepage,?*eh*).
View-2:zju:emp_project(?*en*,?*an*):-($?y_1$,rdf:type,foaf:Person),($?y_1$,foaf:name,?*en*),($?y_1$,foaf:currentProject,$?y_2$),
($?y_2$,rdf:type,foaf:Project),($?y_2$,foaf:name,?*pn*).
View-3:zju:project(?*an*,?*ah*):-($?y_2$,rdf:type,foaf:Project),($?y_2$,foaf:name,?*pn*),($?y_2$,foaf:homepage,?*ph*).
View-4:w3c:emp(?*en*,?*em*,?*an*,?*ah*):-($?y_1$,rdf:type,foaf:Person),($?y_1$,foaf:name,?*en*),($?y_1$,foaf:mbox,?*em*),
($?y_1$,foaf:currentProject,$?y_2$)($?y_2$,rdf:type,foaf:Project),($?y_2$,foaf:name,?*pn*),($?y_2$,foaf:homepage,?*ph*).

Figure 11.3 Mappings from two relational databases with different structures to an ontology. "?*en*,?*em*,?*eh*,?*pn*,?*ph*" are variables and represent "employee name," "employee email," "employee homepage," "project name," and "project homepage," respectively.

TCM case, new databases are regularly added so the total number of databases is increasing gradually. Therefore, the LAV approach is employed in our system, i.e., each relational table is defined as a view of the ontologies. We call such a kind of view a Semantic View.

The lower part of Figure 11.3 shows how to represent the mappings as semantic views in a Datalog-like syntax. As in conventional data integration, a typical semantic view consists of two parts. The left part is called the view head and is a relational predicate. The right part is called the view body and is a set of resource description framework (RDF) triples. There are two kinds of variables in the view definitions. Those variables such as "?*en*,?*em*,?*eh*,?*pn*,?*ph*" are called distinguished variables, which will be assigned by data or instance values. Those variables such as "$?y_1,?y_2$" are called existential variables.

In general, the body can be viewed as a query of the ontology, and it defines the semantics of the relational predicate from the perspective of the ontology. The meaning of semantic view would be clearer if we construct a Target Instance based on the semantic mapping specified by these views. For example, given a relational tuple as below, applying the View-4 in Figure 11.3 on this tuple will yield a set of RDF triples.

Algorithm 11.1 Relational Tuple:

```
w3c:emp("DanBrickley","danbri@w3.org",
    "SWAD","http://swad.org","EU");
```

```
Yielded RDF triples by Applying View-4:
    _:bn1 rdf:type foaf:Person;
foaf:name "Dan Brickley";
foaf:mbox "danbri@w3.org";
foaf:currentProject _:bn2.
_:bn2 rdf:type foaf:Project;
foaf:name "SWAD";
foaf:homepage "http://swad.org".
```

One of the key notions is the newly generated blank node ID. As illustrated above, corresponding to each existential variable $?y$ in the view, a new blank node ID is generated. For example, $:bn_1$, $:bn_2$ are both newly generated blank node IDs corresponding to the variables $?y_1$, $?y_2$ in View-4, respectively. This treatment of an existential variable is in accordance with the RDF semantics, as blank nodes can be viewed as existential variables. We give the formal definition of the semantic view as below. More detailed aspects about the semantic view can be found in Ref. [3].

Definition 11.1. (Semantic View). Let *Var* be a set of variable names. A typical SemanticView is like the form: $R(^{-}X)\!: - \,G(^{-}X,\ ^{-}Y)$, where:

1. $R(^{-}X)$ is called the head of the view, and R is a relational predicate;
2. $G(^{-}X,\ ^{-}Y)$ is called the body of the view, and G is a RDF graph with some nodes replaced by variables in *Var*;
3. The ^{-}X, ^{-}Y contain either variables or constants. The variables in ^{-}X are called distinguished variables, and the variables in ^{-}Y are called existential variables.

11.3.2 Visualized Semantic Mapping Tool

The task of defining semantic mappings from relational schema to ontologies is burdensome and erroneous. Although we could develop tools to automate this process, it still cannot be fully automated and requires human involvement, especially for integration of databases with different schema structures.

Figure 11.4 displays the visualized mapping tool we developed to facilitate the task of defining semantic views. It has five panels. The DBR (Database Record) panel displays the relational schemas, and the OntoSchem panel displays the shared ontology. The Mapping Panel visually displays the mappings from relational schemas to ontologies. Typically, the user drags tables or columns from the DBRes panel, and drags classes or properties from the OntoSchem panel, then drops them into the mapping panel to establish the mappings. By simple drag-and-drop operations, users could easily specify which classes should be mapped in a table and which property should be mapped in a table column. After these operations, the tool automatically generates a registration entry, which is submitted to the semantic registration service. Also, the user could use the Outline panel to browse and query previously defined mapping information, and use the Properties panel to specify some global information, such as namespace, or view the meta-information about the table.

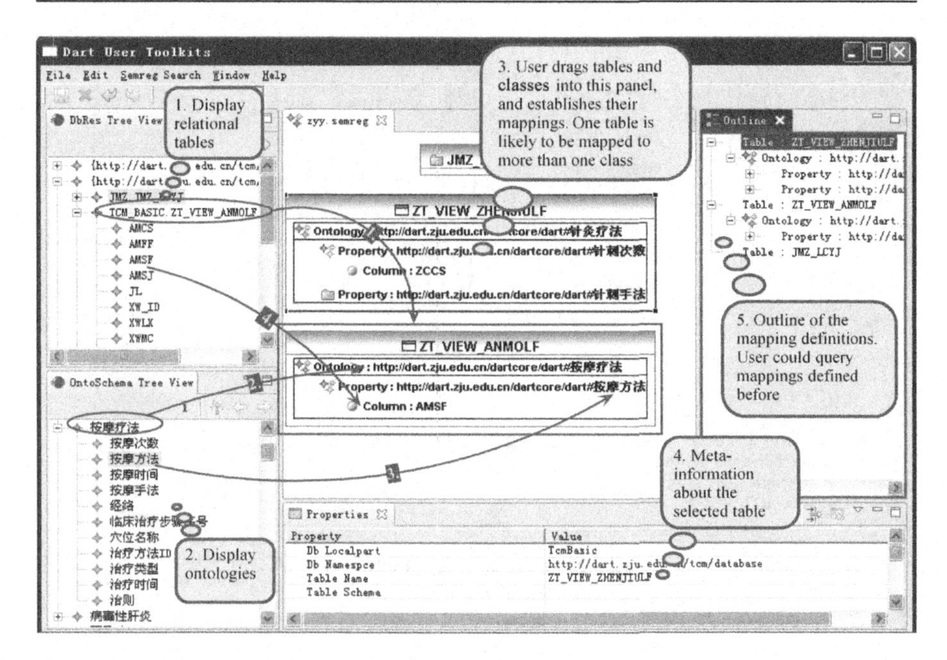

Figure 11.4 Visualized semantic mapping tool.

11.4 TCM Semantic Portals

The semantic mediator is designed to separate data providers and data consumers so that they only need to interact with the semantic layer. For example, developers could write applications using the shared ontology without the need for any knowledge about databases. Besides that, our system also offers two different kinds of user interfaces to support query and search services.

11.4.1 Dynamic Semantic Query Interface

This form-like query interface is intended to facilitate users in constructing semantic queries. The query form is automatically generated according to class definitions. This design provides the extensibility of the whole system—when ontology is updated with the changes in database schema, the interface can dynamically adapt to the updated shared ontology.

Figure 11.5 shows how a TCM user constructs a semantic query. Starting from the ontology view panel on the left, a user can browse the ontology tree and select the classes of interest. A query form corresponding to the property definitions of the selected class will be automatically generated and displayed in the middle. Then a user can check and select the properties of interests or input query constraints into the textboxes. Accordingly, a SPARQL query is constructed and could be submitted to the semantic query service, where the query will be rewritten in a set of SQL

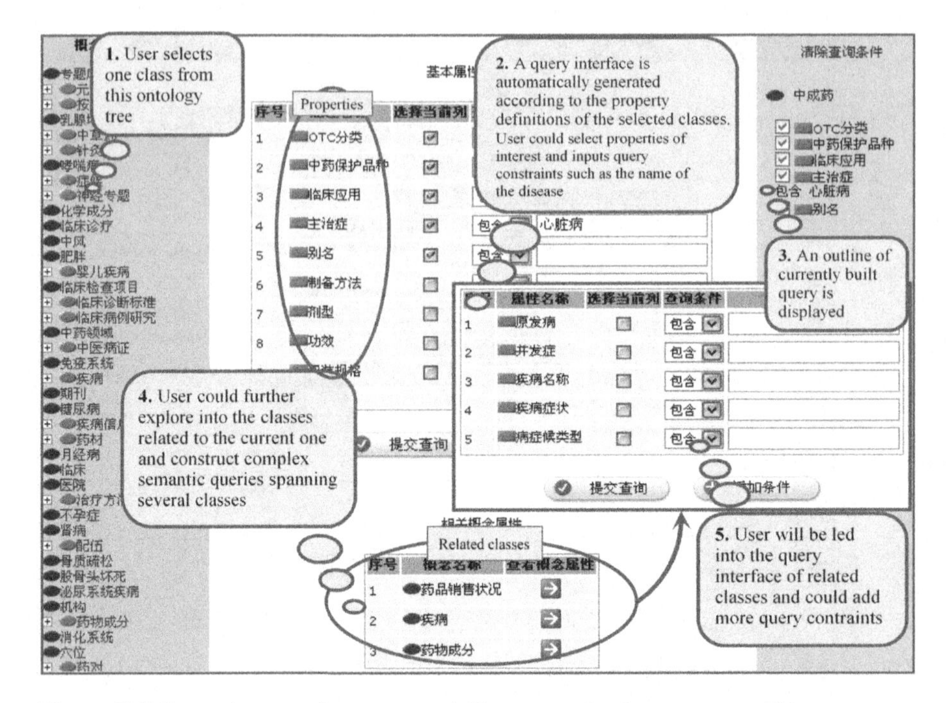

Figure 11.5 Dynamic semantic query portal. Please note that because many Chinese medical terminologies are only available in the Chinese language and they are not always interpretable, we have annotated all the necessary parts of the figures in English.

queries using mapping views contained in the semantic registration service. The query rewriting is a somewhat complicated process, and Ref. [3] gives the detailed introduction to the rewriting algorithm. In addition, a user could define more complex queries. For example, depicted in the lower-middle part of Figure 11.5, a user could follow the links leading to related classes of the current class, and select more properties or input new query constraints.

Figure 11.6 shows the situation in which a TCM user is navigating the query results. Starting with selecting one result highlighted, the user can find out all of the related data entries by following the semantic links. Please note that in this example the relations between the search results and those "discovered" by following the semantic links are derived from the semantic layer.

11.4.2 *Intuitive Search Interface with Concepts Ranking and Semantic Navigation*

Unlike the semantic query interface, this Google-like search interface just accepts one or more keywords and makes a complete full-text search in all databases. Figure 11.7 shows the situation where a TCM user performs some search

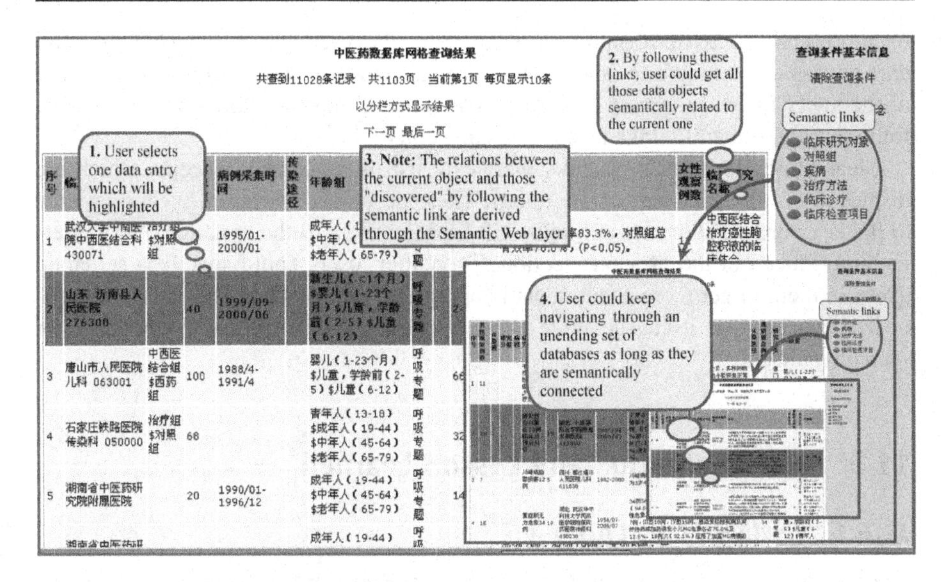

Figure 11.6 Semantic navigation through the query results.

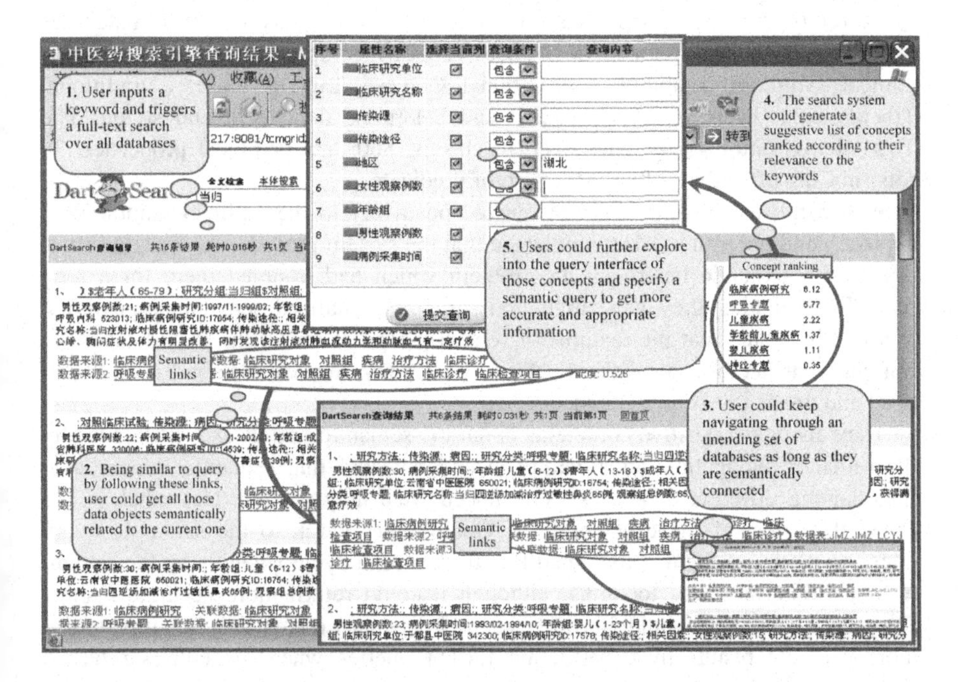

Figure 11.7 Intuitive search portal with concept ranking and semantic navigation.

operations. Starting from inputting a keyword, the user can retrieve all of those data entries containing one or more hits of that keyword. Being similar to the case of the query interface, a user could also semantically navigate the search results by following the semantic links listed with each entry.

Meanwhile, the search system generates a list of suggested concepts which are displayed on the right part of the portal. They are ranked based on their relevance to the keywords. These concept links will lead the users to the dynamic query interface introduced in the previous section. Thereafter, users could specify a semantic query in them to get more accurate and appropriate information. We call it an intuitive search because it can generate a list of suggestions for users to obtain better search results.

11.5 User Evaluation and Lesson Learned

11.5.1 Feedback from CATCM

The first proof-of-concepts prototype was deployed during the fall of 2004. By using that prototype, we convinced our CATCM partner to take the semantic web technologies to help them in managing their fast-increasing TCM databases. After a thorough requirements analysis and with a careful redesign and re-engineering of the entire system, a more stable and user-friendly version was released in September 2005 and deployed at CATCM for open evaluation and real use.

Currently, the system deployed at CATCM provides access to over 70 databases including TCM herbal medicine databases, TCM compound formula databases, clinical symptom databases, a traditional Chinese drug database, a traditional Tibetan drug database, TCM product and enterprise databases, and so on. The TCM shared ontology includes over 70 classes, 800 data or object properties, 30 rdfs:subClassof, and 40 rdfs:subPropertyof predicates.

In general, users from CATCM reacted positively to the entire Semantic Web approach and our system. They indicated that the system provided an amazing solution to the semantic heterogeneity problem which had troubled them for a long time. In particular, they gave high praise to the visualized semantic registration tool and indicated that the features of semantic registration of the new database at runtime considerably saved them a lot of time when the new database was developed and needed to be integrated.

They also gave positive comments about the semantic portals as well, especially the semantic navigation functionality. They indicated that semantic interconnections among different databases were indeed what they wanted. Nevertheless, we found that most of the users prefer a Google-like search to the semantic query interface. Some of them complained that the learning cycle when using the semantic query interface was too long, although it could return more accurate results. They also said they would like to use the concepts ranking functionality to get more accurate results by constructing further queries when the entries returned from the search were overwhelming.

Table 11.1 Results for the Survey of Predicates Usage

Predicates	E1	E2	E3	E4	E5	E6	E7	E8	E9	E10	Average
rdf:datatype	9	10	8	9	10	10	9	7	10	9	9.1
rdfs:subClassOf	8	8	7	9	9	8	8	9	10	7	8.3
rdfs:subPropertyOf	8	8	8	7	8	8	9	9	9	8	8.2
owl:inverseOf	8	8	7	8	7	9	8	9	7	9	8.0
owl:cardinality	7	8	7	7	6	7	9	7	7	9	7.4

11.5.2 A Survey on the Usage of RDF/OWL Predicates

RDF/OWL has offered us a range of predicates, but not all of them are useful for relational data integration. We have made a survey of the usage of RDF/OWL predicates for relational database integration, and the results are indicated in Table 11.1.

In this survey, we invited 10 developers who are familiar with both Semantic Web technologies and our system. Two of them are developers from CATCM. They were asked the same questions: "From a practical view, what do you think are the most important constructs for relational data integration in the Semantic Web," and were requested to write down some explanation for the reason for their choice. We summarized their comments and the score result as follows.

Data-type support was considered to be important, because most commercial RDBMS (Relational Data Base Management System) has a well-defined and unique data-type system. RDFS predicates rdfs:subClassOf and rdfs:subPropertyOf have higher scores because they could enhance the query processing with additional inference capabilities. OWL predicate owl:inverseOf is useful when defining relations in both directions, which is a usual case in relation database integration. One of the developers indicated that predicate owl:inverseOf could help to find more efficient rewritings in some cases. Predicate owl:cardinality is useful in adding more constraints to ensure data integrity.

Some other predicates considered as useful include owl:TransitiveProperty, owl:SymmetricProperty, owl:DatatypeProperty, and owl:ObjectProperty. Some thought both owl:TransitiveProperty and owl:SymmetricProperty could add additional deductive capabilities on top to yield more query results. owl:DatatypeProperty and owl:ObjectProperty could be used to distinguish a simple data value column and foreign key column.

11.6 Related Work

11.6.1 Semantic Web Context

In the Semantic Web community, semantic data integration always has been a noticeable research topic. In particular, there have been a number of works dealing with how to make contents of existing or legacy databases available for Semantic Web applications. A typical one is D2RQ. D2RQ is a declarative language to describe mappings between relational database schemata and OWL/RDFS ontologies and is implemented as a Jena plugin that rewrites RDQL queries into SQL queries. The result sets of these SQL queries are transformed into RDF triples that are passed up to the higher layers of the Jena framework. RDF Gateway is a commercial software having similar functionalities. It connects legacy database resources to the Semantic Web via its SQL Data Service Interface. The SQL Data Service translates an RDF-based query to a SQL query and returns the results as RDF data. Our system is different from D2RQ and RDF Gateway. We take the view-based mapping approach which has a sound theoretical foundation, and we have a visualized mapping tool and ontology-based query and search tool which are not offered by these two systems.

Some other studies propose the direct manipulation of relational data to RDF/OWL format, and then the data could be processed by OWL reasoners or be integrated by an ontological mapping tool. D2RMap, KAON REVERSE, and many other toolkits offer such a kind of reverse engineering functionality. de Laborda and Conrad [4] proposed an ontology called "Relation OWL" to describe the relational schema as OWL, and then use this OWL representation to transform relational data items into RDF/OWL and provide a query service by RDQL. The shortcoming of this kind of approach is that they have to dump all the relational data into RDF/OWL format before querying, which would be impractical if the RDBMS contains huge volumes of data. Moreover, they did not consider the issue of integrating heterogeneous databases using formal ontologies, which is one of the focuses of our solution.

An et al. [5] presented an interesting paper concerned about defining semantic mappings between relational tables and ontologies within the Semantic Web context. They introduced a tool which could automatically infer the LAV mapping formulae from simple predicate correspondences between relational schema and formal ontologies. Although a completely automatic approach to defining semantic mapping is difficult, it would be a great enhancement to our visualized tool if some candidate mapping suggestions could be provided beforehand. That will be one of our future works.

The Drug Ontology Project for Elsevier (DOPE) [6] explores ways to provide access to multiple life science information sources through a single semantic interface called DOPE browser. However, it is still a document management system, mainly concerning a thesaurus-based search, RDF-based querying, and concept-based visualization of large online document repositories. It cannot answer semantic queries such as "What diseases does this drug treat?" or "What kind of drugs can treat this disease?" We have seen that the authors of DOPE are considering it as one of their future studies.

Piazza [7] is an interesting P2P-based data integration system with consideration of Semantic Web vision. But the current system has been implemented with the XML data model for its mapping language and query answering. However, we think P2P architecture would be a promising direction, and we are considering extending our system to support the P2P working mode and test its scalability and usability.

Dou et al. [8] proposed an ontology-based framework called OntoGrate. It can automatically transform relational schema into ontological representation, and users can define the mappings at the ontological level using bridge axioms. Goasdoue [9] considers the theoretical aspect of answering query using views for the Semantic Web, and Haase and Motik [10] introduced a mapping system for OWL-DL (Description Logic) ontology integration.

11.6.2 Conventional Data Integration Context

Without considering the Semantic Web technologies, our solution can be categorized as the topic of "answering query using view," which has been extensively studied in the database community [2,11]. Most previous work has been focused on the relational case [2] and XML case [12].

On the one hand, we believe it would be valuable for the Semantic Web community to take more consideration of the techniques that have been well studied in the database community such as answering query using view. On the other hand, we think that Semantic Web research does raise a lot of new issues and challenges for database researchers. From our experience, the challenges include the following: How to rank the data object just like the page ranking of Google? How to maintain highly evolvable and changeable schema mappings among a great number of open-ended sets of databases with no centralized control?

Moreover, a lot of work has been done in the area of ontology-based data integration [13]. Much of this took some ontological formalism such as DL to mediate heterogeneous databases, and used the view-based mapping approach. In comparison, we implement RDF/OWL-based relational data integration with a Semantic Web vision in mind.

11.7 Conclusions

In this chapter, we presented an in-use application of traditional Chinese medicine enhanced by a range of Semantic Web technologies, including RDF/OWL semantics and reasoning tools. The ultimate goal of this system is to realize the "web of structured data" vision by semantically interconnecting legacy databases that allows a person, or a machine, to start in one database and then move around in an unending set of databases which are connected by rich semantics. To achieve this demanding goal, a set of convenient tools was developed, such as a visualized semantic mapping tool, a dynamic semantic query tool, and an intuitive search tool with concepts ranking. Domain users from CATCM indicated that the system provided an amazing solution for the semantic heterogeneity problem that had troubled them for a long time.

Currently, although this project is complete, several updated functionalities are still under our consideration. To be specific, we are going to enhance the mapping tools with some heuristic rules to automate the mapping task as far as possible, just like the approach proposed by An et al. [5]. Otherwise, we will develop a more sophisticated mechanism to rank the data objects just like the page rank technology provided by popular search engines.

References

[1] Buitelaar P, Olejnik D, Sintek M. OntoLT: a protégé plug-in for ontology extraction from text. In: Proceedings of the international semantic web conference (ISWC); 2003.
[2] Halevy AY. Answering queries using views: a survey. VLDB J 2001;10:270–94.
[3] Chen H, Wu Z, Wang H, et al. RDF/RDFS-based relational database integration. In: ICDE 2006; 2006. p. 94.
[4] de Laborda CP, Conrad S. Bringing relational data into the semantic web using sparql and relational owl. In: International workshop on semantic web and database at ICDE 2006; 2006. p. 55–60.
[5] An Y, Borgida A, Mylopoulos J. Inferring complex semantic mappings between relational tables and ontologies from simple correspondences. In: International semantic web conference; 2005. p. 6–20.
[6] Stuckenschmidt H, van Harmelen F, de Waard A, et al. Exploring large document repositories with RDF technology: the dope project. IEEE Intell Syst 2004;19:34–40.
[7] Halevy AY, Ives ZG, Madhavan J, et al. The piazza peer data management system. IEEE Trans Knowl Data Eng 2004;16(7):787–98.
[8] Dou D, LePendu P, Kim S, et al. Integrating databases into the semantic web through an ontology-based framework. In: International workshop on semantic web and databaseat ICDE 2006; 2006. p. 33–50.
[9] Goasdoue F. Answering queries using views: a KRDB perspective for the semantic web. ACM Trans Internet Technol 2003;4:1–22.
[10] Haase P, Motik B. A Mapping system for the integration of OWL-DL ontologies. In: IHIS'05: Proceedings of the first international workshop on interoperability of heterogeneous information systems; 2005. p. 9–16.
[11] Abiteboul S, Duschka OM. Complexity of answering queries using materialized views. In: The seventeenth ACM SIGACT-SIGMOD-SIGART symposium on principles of database systems; 1998. p. 254–63.
[12] Yu C, Popa L. Constraint-based XML query rewriting for data integration. In: 2004 ACMSIGMOD international conference on management of data; 2004. p. 371–82.
[13] Wache H, Voegele T, Visser U, et al. Ontology-based integration of information—a survey of existing approaches. In: Stuckenschmidt H, editor. IJCAI01 workshop: ontologies and information sharing; 2001. p. 108–17.

12 Probabilistic Semantic Relationship Discovery from Traditional Chinese Medical Literature

12.1 Background

The Semantic Web was first introduced by Tim Berners-Lee, the founder of the World Wide Web, with the aim of building a machine-understandable web. It provides a number of new web languages and technologies such as the resource description framework (RDF) and web ontology language (OWL), which can be used to model domain concepts and describe semantic relations. It is valuable in developing an automatic approach in discovering semantic relationships from textual documents, as the discovered relationships can be used to provide recommendations to ontology engineers, verify known knowledge, and reveal novel insights.

It has been always a significant challenge to discover semantic relationships from textual documents [1], particularly in the domain of traditional Chinese medicine (TCM). One principal reason is that large amounts of TCM knowledge are written in a classical Chinese literary style, which is largely different from Vernacular Chinese and cannot be processed efficiently by existing linguistic analysis methods [2]. Another challenge is to deal with uncertainty for those discovered new semantic relations. The traditional RDF model can only describe and deal with deterministic semantic relationships, whereas the relationships discovered from textual documents are normally of an uncertain nature and need a probabilistic model to store the uncertain information.

In this chapter, we propose a probability-based semantic relation discovery method, which is based on TCM domain ontology and plenty of TCM literature. The TCM ontology is a huge conceptual knowledge system that is composed of two basic components: concept system and semantic system. TCM ontology is integrated into the traditional association rule learning method to guide relationship extraction from textual literature. Association rule learning is used for discovering interesting relations between variables in large datasets. A probability that combines both the weight calculated from textual analysis and the predicted probability in domain ontology is used to verify the certainty weight of candidate semantic relationships. A probability-based RDF model is used to represent the result, in

Modern Computational Approaches To Traditional Chinese Medicine. DOI: http://dx.doi.org/10.1016/B978-0-12-398510-1.00012-1

which the probability information will be preserved along with semantic relations in RDF.

The main contributions to this chapter are twofold. First, we propose a novel method which combines the information in both domain ontology and domain publications to discover and evaluate semantic relationships. Second, we use a probability-based model to construct probability-based semantic relationships, in which the certainty of each semantic relationship will be stored and preserved in RDF format.

12.2 Related Work

Jiang et al. [3] proposed a generalized associations mining method called GP-close. An ontology-based model is used to extract RDF statements from textual web contents, and linguistic analysis techniques are used to process hierarchical information of all statements to produce generalized patterns. The NLP (Natural Language Processing) techniques they used heavily depend on syntax analysis algorithms, which have a very limited application range. Furthermore, the generalized pattern mining method is based on the probability information that is not preserved in the final result. In this chapter, we propose a probability-based semantic association discovery method which can be used to extract semantic associations from textual documents without syntax analysis, and a probability-based RDF model is used to preserve the probability information of results.

de Keijzer and van Keulen [4] introduced a probabilistic RDBMS (Relational Database Management System) approach to solve the problem of automatic data integration. A probabilistic XML (eXtended Markup Language) DBMS (Database Management System) and its relevant operations were introduced by van Keulen and de Keijzer [5], which is used to solve the problem of XML data integration. Both of these approaches concentrated on dealing with the problem of data integration. In this chapter, we bring probabilities into the RDF model to preserve the uncertainty in the semantic association discovery process.

MapFace [6] is a semi-automatic semantic relation verification system based on MetaMap [7] and UMLS. This system was developed by the National Library of Medicine to map biomedical texts with UMLS metathesaurus. It provides a visualized interface to help users extract inherent semantic relationships from biomedical publications. Its relation discovery method is based on individual sentences, ignoring associations implied in more than one sentence. Because it depends on users to verify candidate semantic relations, it becomes inefficient, especially in the case of large-scale data.

Hotho et al. [8] introduced a semantic discovery method using semantic predictions to enhance the literature-based discovery (LBD) system. The predictions, which are produced by two NLP systems, are coupled with an LBD system BITOLA to enhance their capability in new association discovery. Their method is based on the co-occurrence of semantic relations, and its purpose is to improve semantic relations extraction results rather than to discover them.

Ding and Peng [9] proposed a probabilistic extension to ontology language OWL to support uncertain ontology representation and ontology reasoning and mapping.

In their approach, Bayesian networks, a widely used graphic model for knowledge representation under uncertainty, is incorporated with OWL. Fukushige [10] proposed a vocabulary for representing probabilistic knowledge in RDF to present a framework for probability calculation using RDF and a Bayesian network.

12.3 Methods

We use an ontology-based association rule learning method to extract important instances and semantic relations from tens of thousands of examples of TCM domain literature. The candidate semantic relations are associated with a certainty weight that is based on a probability derived from information contained in both textual documents and domain ontologies. Generally speaking, the method comprises four steps: (1) instance extraction, (2) instance pair discovery, (3) semantic relation evaluation, and (4) probability-based semantic relation extraction.

12.3.1 Instance Extraction

The text-processing step first translates raw texts into sets of words to which the association rule learning method will be applied to extract candidate instance pairs. Second, we use a domain dictionary to translate separated sentences into word vectors, which are used as the data source in the association rule learning step.

A simple strategy is used to segment a textual document into sentences. We just scan through the text, extract a sentence when confronting a sentence separator, such as period and exclamation mark. We then use a domain dictionary to extract words from the sentence set. The matched words in the same sentence will be put into a word vector. As the goal is to discover instances and their relations, the words in the domain dictionary either must be directly associated with domain instances or can be mapped to them. It means that for any words in the domain dictionary, there should be a corresponding instance in the domain ontology. One simple strategy of building such a domain dictionary is to use the instance names in the domain ontology, which is very effective when the instance numbers are large. When words are mapped to ontology instances, the word vectors are then translated into instance vectors.

We use $K = \{V_1, V_2, \ldots, V_m\}$ to represent the set of instance vectors, where $V_k = (i_1, i_2, \ldots, i_{kn})$ is the instance vector of sentence i, m is the number of instance vectors which equals the number of sentences, k_n is the number of instances in instance vector k, and i_j $(0 < j < k_n)$ represents the jth matched instance in sentence k. As we can see, the instance vectors are subsets of all the instances in the domain ontology, which we called O. We can then employ the association rule learning method to extract instance relations from those instance vectors.

12.3.2 Instance Pair Discovery

The association rule mining technique was first introduced by Agrawal et al. [11], which was used to discover regularities between products in large-scale transaction data of supermarkets. In our approach, the instances in the domain ontology can be seen as the set of all items. The instance vectors extracted from domain literature can be viewed as transactions.

Generally, our approach depends on a hypothesis that any instances that appear in the same sentence are likely to have a certain kind of relationship. It is obvious that in many languages, words in the same sentence are much more likely to be related than words in different sentences. Based on this hypothesis, we can use the association rule learning method to extract interesting instance pairs from instance vectors.

The primary advantage of our approach is that it has very high recall because all possible combinations of instances in the same sentences will be considered as candidate relations. However, lots of noisy data will be also introduced in the result, too, because those irrelevant instances which just happen to be in the same sentence also will be considered as candidate pairs. Based on the fact that important instance pairs are likely to appear many more times than trivial ones, we use a combination of the association rule learning method and a threshold to decrease the noise. The formula to calculate the weight is defined as follows:

$$w(i_a, i_b) = \sum_{i=1}^{k} c_i f_i(i_a, i_b) \quad \left(\sum_{i=1}^{k} c_i = 1, \ \ c_i > 0 \right) \tag{12.1}$$

where c_i is a constant value which is used to adjust the weights of different association rule learning measures. $f_i(i_a, i_b)$ is the ith association rule measure to evaluate the interestingness of instance pair (i_a, i_b), which is normalized to [0,1]. For example, the $f_{\text{supp}}(i_a, i_b)$ of supp(i_a, i_b) with upper limit a and lower limit b will be:

$$f(i_a, i_b) = \frac{\text{supp}(i_a, i_b)}{b - a} \quad (b > a > 0) \tag{12.2}$$

As we can see, the value of the supp(i_a, i_b) is in $[a, b]$. This formula normalizes the value of supp(i_a, i_b) to [0,1]. Because different measures have different values, the normalization formula also will be different. For example, the supp(i_a, i_b) are always smaller than 1, but lift(i_a, i_b) is likely to be much more larger. In fact, it does not have an upper limit. Therefore, different measures must be mapped to the same value domain. Then they can be used together to calculate the weight of given instance pair.

In the weight computation formula, we use the linear combination of different association rule learning measures to evaluate the interestingness of given instance pairs, which can avoid the drawbacks of using just one specific measure. The weights of different measures can be adjusted by the constant value c_k. We use a threshold to extract interesting pairs from instances vectors, and the value of w(i_a, i_b) will be used to calculate the probability of semantic relations for those pairs, which will be introduced in the Section 12.3.3. Figure 12.1 displays the semantic relationship evaluation.

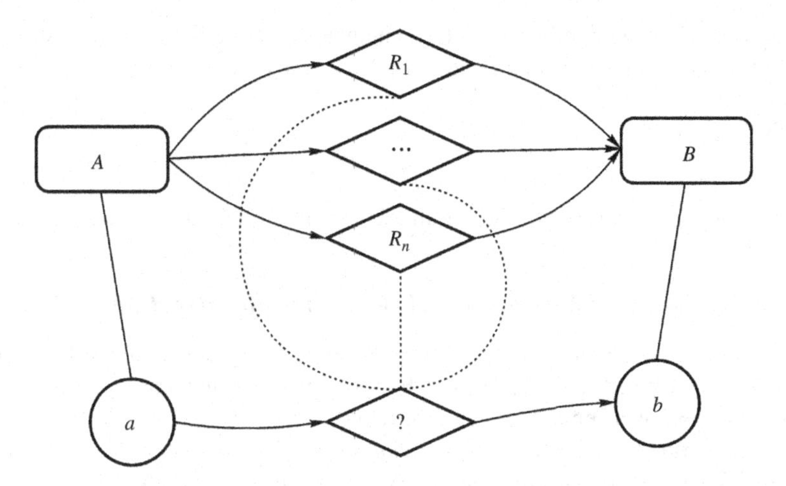

Figure 12.1 Semantic relationship evaluation.

12.3.3 Semantic Relationship Evaluation

In the previous section, we used text-processing techniques and association rule learning methods to extract interesting instance pairs from domain literature. The candidate instance pairs are merely pairs with weights indicating how "important" the relations are. There is no information revealed about the specific type or semantics of those relations.

In this section, we use domain ontology to identify the semantic type of candidate instance pairs. The idea is to use the distribution of existing semantic relations in the domain ontology to estimate possible types of a given instance pair. This process can be further divided into two steps: Step 1 is to use all the possible semantic relations in the domain ontology as candidate semantic relation types, and in Step 2, we use the distribution of these semantic relation types in domain ontology to calculate the possibility for each candidate semantic relation to be the type of a given instance pair.

For a candidate instance pair (i_a, i_b), where i_a belongs to concept A and i_b belongs to concept B, we use $R_{A,B} = \{R_1, R_2, \ldots, R_n\}$ to represent all semantic relations existing in the ontology between concept A and B, where n is the number of semantic relations. The formula that we use to calculate the probability is defined as follows:

$$\mathrm{pred}(i_a, i_b, R_i) = \frac{\mathrm{count}(A, B, R_i)}{\sum_{k=1}^{n} \mathrm{count}(A, B, R_k)} \tag{12.3}$$

where

$$i_a \in A, \quad i_b \in B, \quad 0 < i < n$$

count(A, B, R_i) is the number of all the statements existing between all instances of concept A and B with semantic relation R_i. We can easily see that,

$$\sum_{i=1}^{n} \mathrm{pred}(i_a, i_b, R_i) = 1$$

It means that $R_{A,B}$ is a partition of all the semantic relations between A and B.

12.3.4 Probability-Based Semantic Relationship Extraction

In traditional RDF semantic models, there is no certainty evaluation value that can be associated with semantic relations. For example, a statement (i_a, r, i_b) merely means that there is a semantic relation r between instance i_a and i_b, with no indication of the certainty and degree of the relationship. In the association rule learning method, all the results are calculated by statistical analysis and the certainty factors of all candidate instance pairs are based on probabilities. In the semantic relationship determination section, we use the distribution of existing semantic relations in domain ontology to evaluate semantic relations of the candidate instance pair, which is based on statistical analysis, too. As we can see from Eqs. (12.3) and (12.4), there is likely to be more than one candidate semantic relationship for an instance pair. If we use a deterministic model to describe the interesting semantic relations, all the probability information that may contain very important information will be lost. Because of the inherent probabilistic characteristics of both the methods and the results, a probability-based model is much more suitable for storing our result.

We combine the weight in the association rule learning method and the predicted probability in the semantic relationship determination process to compute the final certainty factor; the formula is as follows:

$$\mathrm{prob}(i_a, i_b, R_i) = \mathrm{pred}(i_a, i_b, R_i) \times w(i_a, i_b) \tag{12.4}$$

where $w(i_a, i_b)$ is the association weight of (i_a, i_b) we calculated in Eq. (12.1), and the pred(i_a, i_b, R_i) is the predicted probability of semantic relation R_i between instance association (i_a, i_b) defined in Eq. (12.3). In Eq. (12.4), we combine the weight calculated from textual documents and the semantic relation probability derived from the domain ontology to calculate the final predicted probability of given instance pairs.

In general, prop(i_a, i_b, R_i) is a probability which can indicate the importance of the candidate semantic relationship R_i for given instance pair (i_a, i_b).

By changing the format of the RDF semantic statement from "triple" to "quadruple," we can store the probability information in a probability-based RDF statement. The set of probability-based RDF statement $P_{a,b}$ for instance pair (i_a, i_b) is defined as follows:

$$P_{a,b} = \{(i_a, R_i, i_b, \mathrm{prob}(i_a, i_b, R_i)) | \mathrm{prob}(i_a, i_b, R_i) > c_{\mathrm{thred}}, R_i \in R\} \tag{12.5}$$

where c_{thred} is the threshold used to extract important semantic associations, R is the set of all candidate semantic relationships between i_a and i_b. As we can see from Eq. (12.5), the quadruple (i_a, R_i, i_b, $p_{a,b}$) means that instance i_a has semantic relationship R_i with instance i_b with the probability $p_{a,b}$.

The final result of semantic associations will be a set of quadruples which contain both the RDF statements and the relevant probabilities, as a set of RDF statements intrinsically can be seen as a labeled, directed multigraph, which represents domain objects and relations between them. With probabilities in the RDF statements, the semantic graph now becomes a weighted graph, from which probability-based semantic query and semantic induction can be performed.

12.3.4.1 Experiments

The experiment is based on a TCM ontology that was developed by the China Academy of Traditional Chinese Medicine. The ontology includes 58 standard semantic relations and 126 standard concept types, with 309,509 instances and 177,984 RDF statements. The large number of instances makes it a good data resource as both a domain dictionary and a domain knowledge base. We use instance names defined in the ontology directly as the domain dictionary that is used for the word segment algorithm. The RDF statements are used as predefined domain knowledge to evaluate the semantic relation type of instance pairs. We used 40,678 abstracts extract from TCM publications for textual analysis. A large number of them are from classical Chinese literature.

A Chinese Word Segmentation system ICTCLAS [2] is used to extract words from these publications. More than 170,000 instance names are extracted from the TCM ontology to be used in the domain dictionary. The 40,678 abstracts are separated into 257,642 sentences, which are further translated into 169,932 valid instance vectors.

We use two association rule learning methods to calculate the weight of given associations, support and lift, which are defined as follows:

$$\text{supp}(i_a, i_b) = \text{supp}(\{i_a, i_b\}) \tag{12.6}$$

$$\text{lift}(i_a, i_b) = \frac{\text{supp}(i_a, i_b)}{\text{supp}(i_a) \times \text{supp}(i_b)} \tag{12.7}$$

and the normalized form of the two measures is

$$\text{supp}_{\text{norm}}(i_a, i_b) = \frac{\text{supp}(i_a, i_b)}{b - a} \tag{12.8}$$

$$\text{lift}_{\text{norm}}(i_a, i_b) = \text{lift}(i_a, i_b) \times \min(\text{supp}(i_a), \text{supp}(i_b)) \tag{12.9}$$

where

$$\text{supp}(a,b)\in[a,b], \quad b>a>0$$

$$\min(x,y)=\begin{cases} x(x<y) \\ y(x<x) \end{cases}$$

where b and a are the upper limit and the lower limit of supp(i_a, i_b), the function min (x, y) return the smaller one of x and y. Because the value of lift (i_a, i_b) lies in [0, min(supp(i_a), supp(i_b))], the lift$_{\text{norm}}$(i_a, i_b) will be in [0,1].

We use [0.003, 0.1] as the value range of support, which means that the valid instance associations have appeared in the same sentences at least 56 times. We set $c_1=0.2$, $c_2=0.8$ as the weights of support and lift, which means the lift is much more important than support in the experiment.

We set the threshold of the final probability to be 0.1. In total, 724 instances pairs are extracted from domain literature, which construct 12,086 RDF statements by using all the candidate semantic relations. By combining the two methods and using a threshold to filtrate trivial relations, we get 524 probability-based semantic relations as the final result.

12.4 Results and Discussions

Table 12.1 illustrates a representative set of the final result. For the sake of demonstrating the result clearly, we have also included the weights calculated in the association rule learning section and the predictions derived in the semantic association discovery section.

For examples, the result 1 (purpura, treated by, blood platelet, 0.33) and result 2 (hemiplegia, caused by, cerebral apoplexy) represent the connotative and meaningful semantic relations extracted from domain publications by our approach. Our approach recommends that purpura can be treated by blood platelet at a possibility of 0.33, which in reality means the lack of blood platelet is an important reason for purpura.

By using the domain ontology and the domain publications, our approach can not only extract these relations from domain publications but also gives the certainty factor of these relations.

Table 12.1 Probability-Based RDF Statements

...	...	...	...	...	...	...
5	High-density lipoprotein cholesterol	Cholesterol	Isa	0.53	0.40	0.21
6	Bellyache	Diarrhea	Caused by	0.17	0.61	0.10
7	Falling sickness	Attack	Diagnosis by	0.20	0.68	0.13
8	Sarcoma	Mouse	Issue in	0.66	0.26	0.17
9	Sarcoma	Mouse	Treated by	0.66	0.68	0.45
10	Ligustrazine	Rat	Hyponym	0.18	0.98	0.17
...	...	...	...	...	...	...

The results 3–6 are semantic relations that can easily be understood by most people. For example, the semantic relation (high-density lipoprotein hyponym, cholesterol, cholesterol, 0.21) and (high-density lipoprotein cholesterol, isa, cholesterol, 0.21) tell us that HDL-C and cholesterol have the same probability for the semantic relationships "isa" and "hyponym." In our opinion, "isa" is a clear choice, but the machine cannot find the right answer without relevant information. So in our approach, both of the candidate relations were preserved along with the probabilities, which can help the user find the correct answer.

The results 7 and 8 represent semantic relations that make some sense, but not much. These relations may be useful as a clue for the domain expert to discover new knowledge. The results 9 and 10 represent the bad results using our approach, which comprise noises that failed to be filtrated.

12.5 Conclusions

In this chapter, we propose a probability-based semantic relation discovery approach, which combines the domain ontology information and domain literature. A probability-based RDF model is developed to preserve semantic relationships along with the relevant probabilities.

A relationship discovery method based on the association rule learning method is introduced, in which we use the domain dictionary and association rules learning method to extract significant instance pairs from raw texts. We then further introduce a semantic relationship discovery approach, which combines the weight in text mining and the predicted probabilities in domain ontology to produce a uniform probability. Then we can verify and extract the semantic relations based on this probability.

We used TCM domain ontology and more than 40,000 relative publications to verify our approach. It produced 524 candidate semantic relationships. The results have been received positively by researchers in the TCM field, and some results are already used to enrich and verify the domain ontology.

References

[1] Stumme G, Hotho A, Berendt B. Semantic Web mining: state of the art and future directions. Web Semantics: Sci Serv Agents World Wide Web 2006;4(2):124–43.

[2] Zhang HP, Yu HK, Xiong DY, et al. HHMM-based Chinese lexical analyzer ICTCLAS. In: Proceedings of the second Sighan workshop affiliated with 41st ACL. Sapporo, Japan; 2003. p. 184–87.

[3] Jiang T, Tan A, Wang K. Mining generalized associations of semantic relations from textual web content. IEE Trans Knowl Data Eng 2007;19(2).

[4] de Keijzer A, van Keulen M. A possible world approach to uncertain relational data. In: SIUFDB-04 international workshop on supporting imprecision and uncertainty in flexible Databases. Zaragoza, Spain; 2004.

[5] van Keulen M, de Keijzer A. A probabilistic XML approach to data integration. In: Proceedings of the 21st international conference on data engineering (ICDE'05); 2005. p. 459–70.

[6] Gschwandther T, Kaiser K, Miksch S. MapFace: a graphical editor to support the semantic annotation of medical text. In: Proceedings of the junior scientist conference; 2008. p. 91–2.

[7] Aronson AR. Effective mapping of biomedical text to the UMLS metathesaurus: the MetaMap program. In: Proceedings of the AMIA symposium. Washington, DC; 2001. p. 17–21.

[8] Hotho A, Nürnberger A, Paaß G. A brief survey of text mining. LDV Forum 2005;20(1):19–62.

[9] Ding ZL, Peng Y. A probabilistic extension to ontology language OWL. Proceedings of the 37th Hawaii international conference on system sciences; 2004.

[10] Fukushige Y. Representing probabilistic knowledge in the emantic Web; 2004. Available from: <http://www.w3.org/2004/09/13-Yoshio/PositionPaper.html>.

[11] Agrawal R, Imielinski T, Swami A. Mining association rules between sets of items in large databases. In: SIGMOD conference; 1993. p. 207–16.

13 Deriving Similarity Graphs from Traditional Chinese Medicine Linked Data on the Semantic Web[1]

13.1 Introduction

Due to the advancement of the Semantic Web, the sizes of linked datasets as well as linked data-based applications are increasing rapidly [1]. Linked data is the data exposed, shared, and connected via URL (Unified Resource Locator) on the web. It uses URLs to identify things such as resources to facilitate people to dereference them. It also provides useful information about these resources, as well as links to other related resources which may improve information discovery [2].

In linked data-based applications, similar resources discovery is one of the most important problems. For example, in a linked dataset of traditional Chinese medicine (TCM), domain experts may frequently need to find herbs significantly similar to a given herb while making a prescription or carrying out other research activities. The primary idea of prior work is essentially based on either ontology or taxonomy or description statements. However, these methods have two main disadvantages. First, most of these methods are not applicable for the linked datasets with different types of relationships. Second, during the similarity calculation between two resource nodes, only their direct correlated nodes are considered and the effect of the whole semantic graph is ignored.

In this chapter, we propose a novel approach for deriving similarity graphs from open linked data with different types of relationships on the Semantic Web. Semantic similarity transition (SST) is proposed to calculate the similarity between two resources. Inspired by the PageRank [3] algorithm, SST takes the whole semantic graph into consideration. It calculates the similarity between two resources using a more accurate method and transits it inside the graph iteratively until the result between each resource pair reaches a steady status. After each iteration, the similarity for each resource pair gets closer to its true similarity value. In

[1] Mi J, Chen H, Lu B, Yu T. Deriving similarity graphs from traditional chinese medicine linked data on semantic web. IEEE international conference on information reuse and integration, 10–12 August; 2009. p. 157–62. Reprinted with permission from IEEE.

Modern Computational Approaches To Traditional Chinese Medicine. DOI: http://dx.doi.org/10.1016/B978-0-12-398510-1.00013-3

addition, we develop a system that enables the smooth interaction and visualization of the derived similarity graph and facilitates experts to conduct medical research easily and intuitively.

SST has been evaluated using the real linked dataset of TCM. The evaluation result shows that it performs better in calculating similarity than existing methods and the system plays an important role in a use case of clinical decision making.

The rest of this chapter is organized as follows. Section 13.2 reviews the most popular similarity measure techniques. The logical method that we propose is described in Section 13.3. Then, in Section 13.4, we summarize the results that we have obtained from the experimental evaluation in TCM. Finally, we conclude this chapter in Section 13.5.

13.2 Related Work

Semantic similarity is an important problem that has been widely discussed in many research areas [4]. There are several approaches that support entity matching from the Semantic Web. Here we present a brief survey of some methods of semantic similarity measurement, as well as different semantic graphs which can be classified into two categories: an ontology taxonomy-based approach and relationship-based approach.

13.2.1 Taxonomy-Based Approach

The taxonomy-based approaches have been proposed mostly for evaluating semantic similarity between concepts in isa hierarchical ontology, such as WordNet and Gene Ontology [5]. As the simplest similarity measures, some of them compute the distance between two concepts. The measure proposed by Resnik [6] is based on the number of edges between two concepts; Rada et al. [7] proposed a metric to measure the conceptual distance; and Wu and Palmer [8] defined the similarity as the ratio between the edge count of a common parent and the sum count of each concept with root.

Other methods use the features of taxonomy. In another paper of Resnik [9], the likelihood of instances of a concept is denoted by the specialization of the concept; Benatallah et al. [10] proposed a metric to measure concept similarity based on description logic (DL); Li and Horrocks [11] and Castillo et al. [12] used DL as the matching algorithm, meanwhile Paoluccdi et al. [13], Said et al. [14], and Ngan et al. [15] used terms such as "exact," "subsumes," "inversesubsumes," and "fails" to define the similarity between concepts.

13.2.2 Relationship-Based Approach

Relationship-based approaches utilize the amount of needed information to state the common and entire information between these two concepts. Lin [16] defined the similarity between two concepts as the ratio between them. Hau et al. [4]

proposed an approach called inference-based information value (IBIV) and regard the ratio between their IBIVs as similarity.

Some other approaches are based on a combination of the above two methods. Shariatmadari et al. [5] combined these two methods by proposing a weighted parameter. Ehrig et al. [17] proposed a framework for measuring similarity, which consists of three layers, including data, ontology, and context.

Each of the existing methods has its strengths and weakness. In summary, according to our practical evaluation result presented in Section 13.4, most of these methods (except Ref. [4]) are not applicable for the linked datasets with different types of relationships. Furthermore, as mentioned in the introduction, these methods consider each resource as an isolated node during similarity calculation and ignore the effect of the whole semantic graph of the linked dataset.

13.3 SST Approach

In the subsequent sections, we briefly present the algorithm of similarity transition and the calculation of similarity between two resources in a linked dataset with different types of relationships.

13.3.1 Similarity Transition

Motivation: a linked dataset is a kind of labeled directed graph cross domain, which is used for knowledge presentation and cognitive model foundation. Each link represents a kind of relationship between two resources, and they can be represented as a statement in resource description framework (RDF). In these statements, because objects are the property value of subjects and describe their features, the similarity between two subjects can be calculated from the similarity between their corresponding sets of objects. If a linked dataset is considered as a whole semantic graph, then the calculated similarity value between two subjects can be further transited to their own related subjects, which may make the similarity between the related subjects more accurate. We call this kind of calculation similarity transition. It is somewhat like the classical PageRank algorithm, according to which the importance of one page is calculated from the importance of all the other pages that link to it.

For example, Figure 13.1 shows an example subset of a linked dataset, in which X is a class of Paper; Y is a class of Author; and (c, e) and (c, d) belong to the object set of a and b, respectively. The similarity between a and b can be calculated from the similarity between their object sets, including (c, e) and (c, d). Then the result is continuously used to calculate the similarity between f and g.

Algorithm of similarity transition: to calculate the similarity between every resource pair[2] in a semantic graph, first an initial arbitrary value (SST uses 0.0, but

[2] A resource pair refers to two resources with the same node type, such as (a, b), (f, g), (c, e), and (c, d) in Figure 13.1.

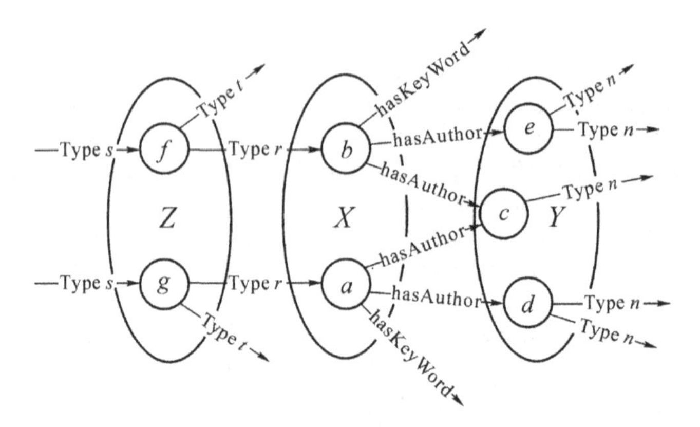

Figure 13.1 An example of resources and links.

the actual value does not make any difference) is assigned to each pair; second, an iterative calculation is executed to enable each value to get closer to the true similarity result by transiting the calculated similarity value inside the semantic graph iteratively. The iterative step stops when every result reaches a steady status. For each iteration, the similarity between each resource pair is calculated from the similarity between their corresponding sets of objects.

13.3.2 Similarity between Sets of Objects

The calculation of similarity between two sets of objects, O_a and O_b, is based on the link types with their subjects. Each set is first divided into several subsets according to the link types, and then each similarity between two subsets with the same link type is calculated separately. Finally, the similarity between O_a and O_b is defined to be a weighted aggregate of the calculated similarity between each subset pair.

Similarity calculation of two sets with the same link type: our definition of similarity between two sets is built upon the information theory-based measure developed in Ref. [10]. Similarity between two sets M and N is defined as the amount of common information, which is formally given by:

$$\text{sim}(M,N) = \frac{f_{\text{common}}(M,N)}{f_{\text{total}}(M,N)} \tag{13.1}$$

where f_{common} is a function measuring the value of common information between M and N, and f_{total} is a function giving the value of the total information. Then the similarity is defined as the ratio of the common information to the total information about these two sets.

By using the inclusion–exclusion principle, the value of total information is given by

$$f_{\text{total}}(M,N) = \text{size}(M) + \text{size}(N) - f_{\text{common}} \tag{13.2}$$

In existing methods, common information is only defined as an intersection of the sets [10] or their inferences [4]. However, the nodes outside the intersection set also may have some similarity by considering the similarity transited from other related nodes to these nodes; consequently, these kinds of nodes also should be taken into consideration. Considering this key aspect, we calculate the common information value between M and N according to

$$f_{\text{common}} = \max\left(\sum_{k=1}^{s} \text{sim}(M_k, N_{i_k})\right) \tag{13.3}$$

where $s = \max(\text{size}(M), \text{size}(N))$; M_k is the kth element of M and $\{i_1, i_2, \ldots, i_s\}$ is one of the permutations of $\{1, 2, \ldots, n\}$.

Similarity weight of each object subset: in a linked dataset with different kinds of relationships (links) between resources and objects, a similarity weight for each subset of objects with a specific link type is assigned according to how important the link is, and it can be adjusted depending on the application context. For example, in Figure 13.1, the similarity weight of object sets with the link type *hasAuthor* should be different from the weight of object sets with the link type *hasKeyWord.*

In summary, the similarity between O_a and O_b is calculated by the following formula:

$$\begin{aligned}\text{sim}_{\text{final}}(O_a, O_b) &= \sum_{i=1}^{m}\sum_{j=1}^{n} \text{weight}_i \times \text{sim}(O_{a_i}, O_{b_j}) \\ &= \sum_{i=1}^{m}\sum_{j=1}^{n} \text{weight}_i \times \frac{\max\left(\sum_{k=1}^{s} \text{sim}(O_{a_{i_k}}, O_{b_{j_k}})\right)}{2 \times \max(\text{size}(O_{a_i}), \text{size}(O_{b_j})) - \max\left(\sum_{k=1}^{s} \text{sim}(O_{a_{i_k}}, O_{b_{j_k}})\right)}\end{aligned} \tag{13.4}$$

where m and n are the number of link types included by O_a and O_b, respectively, $\sum_{i=1}^{m} \text{weight}_i = 1$, O_{a_i} is the subset with the ith link type of O_a, and $\text{sim}(O_{a_i}, O_{b_j}) = 0$ if the corresponding link types are different.

13.4 Experiments and Results

The goal of the experiments is to show that SST can provide high-quality information about the similarity between two resources. It consists of two main parts: the first part compares SST with the most related method to demonstrate that SST provides a better result; the second part briefly shows the visualization of the derived similarity graph.

Because taxonomy-based methods assign the same similarity between every resource pair with the same node type, we will not discuss them here. Among the relationship-based methods, only Ref. [4] has taken different link types into consideration. Therefore, we decide to choose Ref. [4] for comparison.

13.4.1 Dataset Preparation

For testing, a real-world linked data subset is extracted from a large linked dataset of TCM. The structure of its schema is shown in Figure 13.2. It contains 1036 herbs, 413 diseases, 395 formulae, 9 tastes, 6 flavors, and 13 channels. The weight of each object set is assigned by experience based on the link type, as given in Table 13.1.

To compare SST with the method proposed in Ref. [4], we choose to visualize the similarity value between ginseng2 and other herbs, which is calculated using these two methods.

According to domain knowledge, two important factors have been paid attention to before calculating the similarity: (1) the herbs that have a more common curative effect have higher similarity and (2) while experts are looking for herbs

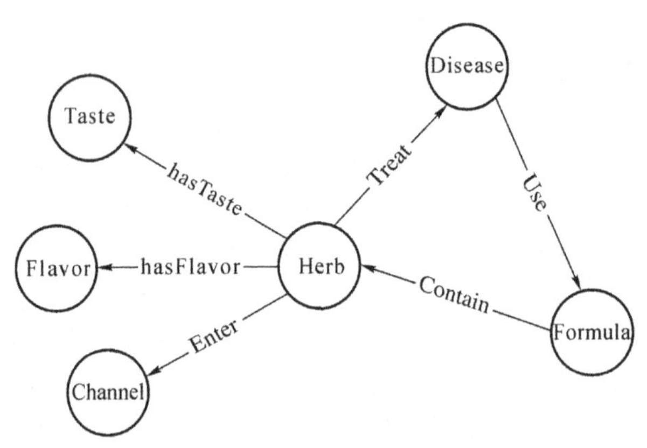

Figure 13.2 Structure of linked data schema.

Table 13.1 Similarity Weight

Link Type	Subject Set	Object Set	Weight
hasTaste	Herb	Taste	0.15
hasFlavor	Herb	Flavor	0.15
Enter	Herb	Channel	0.10
Treat	Herb	Disease	0.60
Contain	Formula	Herb	1.00
Use	Disease	Formula	1.00

significantly similar to a given herb while making a prescription or carrying out other medical research activities, only the herbs that have a common curative effect with the given herb are considered.

Consequently, these herbs are further classified into three categories according to the curative effect of ginseng: (1) closely similar herbs, such as *Aconitum carmichaeli Debx* and "*Ongstamen* Onion Bulb"—both of them and ginseng treat cardiovascular diseases; (2) similar herbs which are used to treat cerebrovascular diseases, such as *Eucommia ulmoides Oliv* and *Curcuma aromatica Salisb*; and (3) dissimilar herbs, such as *Radix Rhizoma Rhei* and *Rhizoma Zingiberis Recens*, which are used to treat other diseases, such as constipation, begma, and asthma.

13.4.2 Results Analysis

Figures 13.3–13.5 show the comparative results of three categories, respectively. In each figure, the selected herbs as well as their curative effects are also listed for an intuitive comparison. Also, in order to show the effects of semantic similarity transition, the result after each time of iteration is shown in Figure 13.3.

As can be seen from these figures, by using SST for the herbs closely similar to ginseng, the calculated similarity is improved significantly; the result is also improved for the herbs similar to ginseng; the similarity between the dissimilar

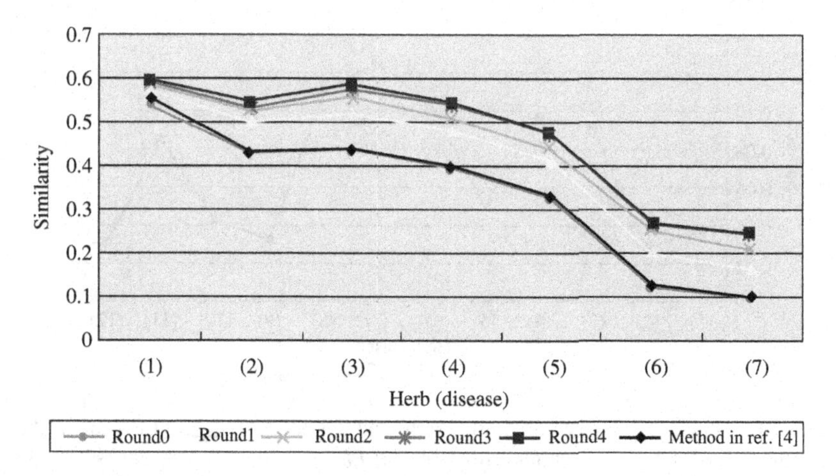

Figure 13.3 Similarity between closely similar herbs and ginseng. (1) *Aconitum carmichaeli Debx* (cor pulmonale and heart failure), (2) Red sage root (coronary heart disease and angina cordis and hyperpiesia), (3) *Longstamen* onion bulb (coronary heart disease and angina cordis), (4) *Radix ophiopogonis* (coronary heart disease and angina cordis), (5) Szechuan Lovage Rhizome (coronary heart disease and angina cordis), (6) *Herba Ecliptae* (coronary heart disease and angina cordis), (7) *Trichosanthes* (coronary heart disease and angina cordis).

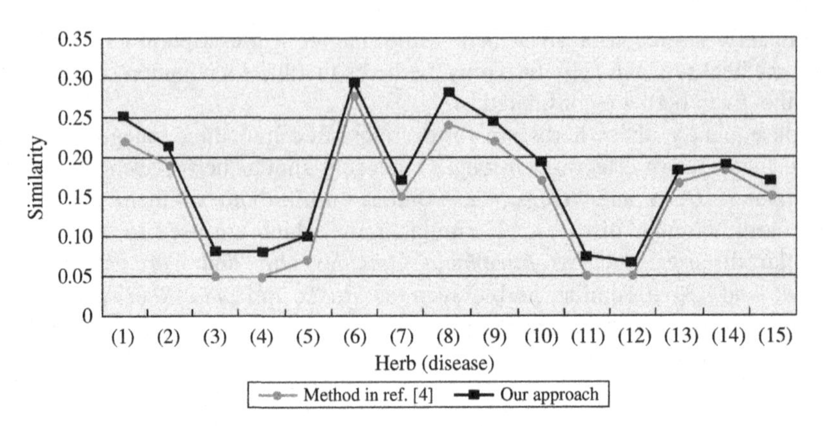

Figure 13.4 Similarity between similar herbs and ginseng. (1) *Eucommia ulmoides* Oliv (hyperpiesia), (2) *Chrysanthemum* (hyperpiesia), (3) Pearl (hyperpiesia), (4) Radices rehmanniae (hyperpiesia), (5) *Curcuma aromatica* Salisb (hyperlipemia), (6) Hawthorn fruit (hyperlipemia and hyperpiesia), (7) Bezoar (cerebral hemorrhage), (8) Chinese *Angelica* (cerebral ischemia and coronary heart disease and angina cordis), (9) Frankincense (cerebral thrombosis), (10) Muskiness (cerebral thrombosis), (11) Myrrh (cerebral thrombosis), (12) *Bombyx batryticatus* (diabetes mellitus and hyperlipemia), (13) Safflower (algomenorrhea and hyperpiesia and cerebral hemorrhage), (14) *Semen persicae* (algomenorrhea and cerebral thrombosis), and (15) *Loranthus yadoriki* (lumbago and hyperpiesia).

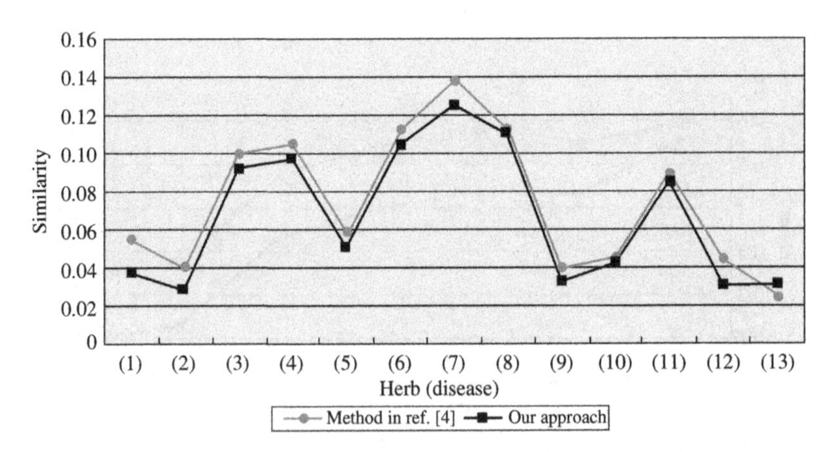

Figure 13.5 Similarity between dissimilar herbs and ginseng. (1) Radix et Rhizoma Rhei (constipation), (2) Glauber's salt (constipation), (3) *Rhizoma zingiberis recens* (subcalorism and begma), (4) Fructus psoraleae (begma), (5) *Gypsum fibrosum* (begma and subcalorism and pneumonia), (6) *Rhizoma Pinelliae* (begma and stagnation of phlegm), (7) Almond (begma and asthma), (8) *Herba ephedrae* (begma and asthma), (9) Bitter Orange (anabrosis and gastritis and constipation), (10) *Grifola* (dropsy), (11) *Euphorbia kansui Liou* (dropsy and fluid retention in the chest and hypochondrium), (12) Chinese Thorowax Root (gastritis and depression), and (13) *Herba Siegesbeckiae* (epilepsia and leucoderma and hyperpiesia).

herbs and ginseng is lower than the result calculated using the existing method, which is more accurate. Furthermore, from Figure 13.3, we can also see that by semantic similarity transition, the result gets better after each iteration until it reaches a stable value.

13.4.3 Result Visualization

The similarity graph is finally derived based on the calculated similarity value. A system is developed to facilitate conducting medical research based on this graph. The screenshot in Figure 13.6 shows a typical result around ginseng. The result has been certified by domain experts.

Several details have been paid attention to during implementation, such as follows: (1) the herbs are positioned in order by the similarity with the user-specified herb; (2) only the herbs most similar to the user-specified herb are retrieved and visualized in the initial view of this graph; and (3) the related resources between user-specified herb and other herbs are indicated to briefly explain why they are similar. For example, in Figure 13.6, cassia twig is the most similar herb to ginseng because it shares much more common information with ginseng than other herbs, such as the formula "Fired *Glycyrrhiza* Decoction," sweet taste, and a warm flavor, for the lung channel and arrhythmia disease.

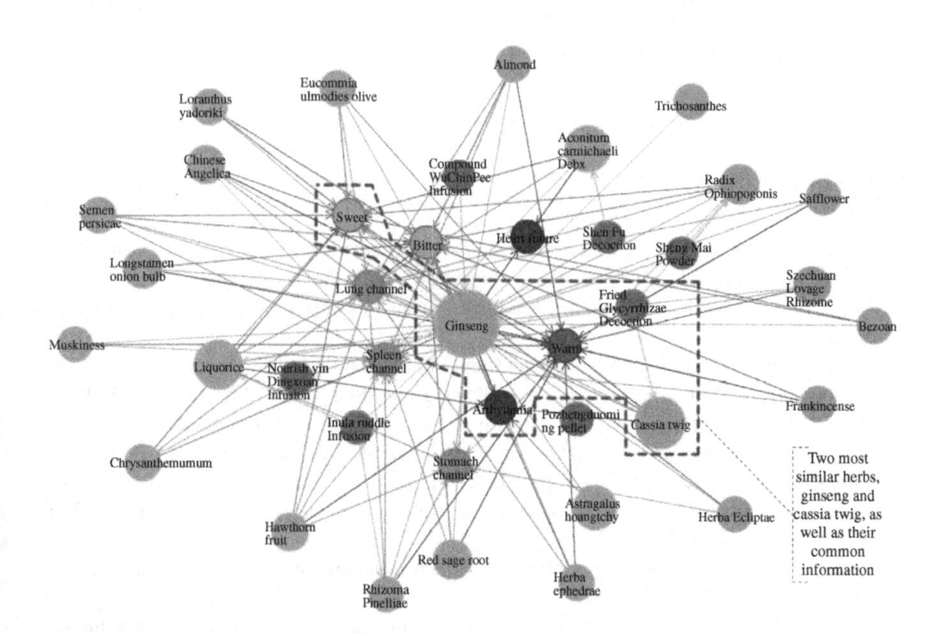

Figure 13.6 Similarity graph around ginseng.

13.5 Conclusions

In this chapter, we presented SST to calculate and improve the similarity between two resources, which consists of three main parts:

1. Similarity transition is proposed to transit the similarity inside the graph by considering the linked data as a whole semantic graph.
2. A comprehensive method is used to calculate the similarity between two sets of objects; the similarity between objects not only in the intersection set but also outside is considered during calculation.
3. A weighted aggregate of similarity is used to improve the final result by considering different link types in the linked dataset.

An empirical evaluation shows that it outperforms the existing method. Furthermore, a system is implemented to enable smooth interaction and visualization of the similarity graph which is derived based on the calculated similarity.

The following work is planned for future research:

1. One of the most important determinants of the result calculated using SST is the topology of the linked dataset. In the future, we will do further research on it to improve the SST algorithm.
2. Currently, the similarity weight of each link type is given from experience. We are developing a machine learning-based algorithm to improve the similarity weight for a better similarity result.
3. The similarity between the resource pair with the same node type is calculated in this chapter. In the future, we intend to focus on the correlation calculation between resources with different node types to derive a correlative graph in a similar way. We believe that the derivation of the correlation graph can assist domain experts in conducting medical research more easily and intuitively.

References

[1] Berners-Lee T, Chen Y, Chilton L, et al. Tabulator: exploring and analyzing linked data on the semantic web. SWUI06 workshop at ISWC06. Athens, Georgia; 2006. p. 22–8.
[2] Berners-Lee T. Linked data. Available from: <http://www.w3.org/DesignIssues/LinkedData>.
[3] Page L, Brin S, Motwani R. The pageRank citation ranking: bringing order to the web, technical report. Stanford university database group; Stanford, California, U.S. 1998.
[4] Hau J, Lee W, Darlington J. A semantic similarity measure for semantic web service. In: Web service semantics workshop at WWW2005. Chiba, Japan; 2005. p. 28–36.
[5] Shariatmadari S, Mamat A, Ibrahim H, et al. Ranking semantic similarity association in semantic web. In: International semantic web conference (posters and demos); 2008. p. 77–9.
[6] Resnik P. Using information content to evaluate semantic similarity. In: Proceedings of the IJCAI05; 1995. p. 488–53.
[7] Rada R, Mili H, Bicknell E. Development and application of a metric on semantic nets. IEEE Trans Syst Man Cybern 1989;19:17–30.

[8] Wu Z, Palmer M. Verbs semantics and lexical selection. In: Proceedings of the 32nd annual meeting on association for computational linguistics, Las Cruces, New Mexico. Morristown, NJ: Association for Computational Linguistics; 1994. p. 133.
[9] Resnik P. Semantic similarity in a taxonomy: an information-based measure and its application to problems of ambiguity in natural language. J Artif Intell Res 1999;11:95–130.
[10] Benatallah B, Hacid MS, Leger A, et al. On automating web services discovery. VLDB J 2005;14(1):84–96.
[11] Li L, Horrocks I. A software framework for matchmaking based on semantic web technology. Twelfth international world wide web conference. Budapest, Hungary: ACM Press; 2003. p. 331, 1-58113-680-3.
[12] Castillo JG, Trastour D, Bartolini C. Description logics for match making of services. Chapter presented at the workshop on application of description logic; 2001.
[13] Paolucci M, Kawamura T, Payne TR, et al. Semantic matching of web services capabilities. First international semantic web conference (ISWC 2002). Sardinia, Italy; 2002. p. 333.
[14] Said MP, Matono A, Kojima I. SPARQL-based OWL-S service matchmaking, SMR 2006. First international workshop on semantic matchmaking and resource retrieval: issues and perspectives. Seoul, Korea; 2006. p. 178 [Online proceedings of CEUR-WS].
[15] Ngan LD, Hang TM, Goh A. Semantic similarity between concepts from different OWL ontologies. 2006 IEEE international conference on industrial informatics. Singapore; 2006.
[16] Lin D. An information-theoretic definition of similarity. In: Proceedings of international conference on machine learning; 1998.
[17] Ehrig M, Haase P, Hefke M, et al. Similarity for ontology—a comprehensive framework. In: Workshop enterprise modeling and ontology: ingredients for interoperability; 2004.

Printed in the United States
By Bookmasters